T0214439

SpringerBriefs in Philosophy

SpringerBriefs present concise summaries of cutting-edge research and practical applications across a wide spectrum of fields. Featuring compact volumes of 50 to 125 pages, the series covers a range of content from professional to academic. Typical topics might include:

- A timely report of state-of-the art analytical techniques
- A bridge between new research results, as published in journal articles, and a contextual literature review
- A snapshot of a hot or emerging topic
- An in-depth case study or clinical example
- A presentation of core concepts that students must understand in order to make independent contributions

SpringerBriefs in Philosophy cover a broad range of philosophical fields including: Philosophy of Science, Logic, Non-Western Thinking and Western Philosophy. We also consider biographies, full or partial, of key thinkers and pioneers.

SpringerBriefs are characterized by fast, global electronic dissemination, standard publishing contracts, standardized manuscript preparation and formatting guidelines, and expedited production schedules. Both solicited and unsolicited manuscripts are considered for publication in the SpringerBriefs in Philosophy series. Potential authors are warmly invited to complete and submit the Briefs Author Proposal form. All projects will be submitted to editorial review by external advisors.

SpringerBriefs are characterized by expedited production schedules with the aim for publication 8 to 12 weeks after acceptance and fast, global electronic dissemination through our online platform SpringerLink. The standard concise author contracts guarantee that

- an individual ISBN is assigned to each manuscript
- each manuscript is copyrighted in the name of the author
- the author retains the right to post the pre-publication version on his/her website or that of his/her institution.

More information about this series at http://www.springer.com/series/10082

C. Mantzavinos

A Dialogue on Explanation

C. Mantzavinos
Department of History and Philosophy of Science
University of Athens
Athens, Greece

ISSN 2211-4548   ISSN 2211-4556   (electronic)
SpringerBriefs in Philosophy
ISBN 978-3-030-05833-3   ISBN 978-3-030-05834-0   (eBook)
https://doi.org/10.1007/978-3-030-05834-0

Library of Congress Control Number: 2018964021

This Springer imprint is published by the registered company Springer Nature Switzerland AG
The registered company address is: Gewerbestrasse 11, 6330 Cham, Switzerland

Γιά τήν Ἕλενα καί τόν Ἀλέξανδρο

Preface

We are all philosophers, and we develop our own philosophy by exchanging views and arguments. The dialogue form is and should remain the principal form of philosophizing, since ideas, like butterflies, do not merely exist—they develop. This is certainly the case in actual philosophical interaction, and it can be the case in written philosophical exposition. However, the dialogical form has been systematically neglected as a way to philosophize. Although there are some exceptions, the essay form has been the dominant form used to express philosophical ideas and arguments in modern philosophy. I see philosophy as an ongoing, open-ended dialogue, and the aim of the book is to help revive this great philosophical tradition and to address philosophical problems in a dialectical form.

I have started writing this dialogue in November 2013 during my short visit at the University Center for Human Values at Princeton and finished the first draft in January 2015 in Athens. I have continued working on it during my visit at the Maison des Sciences de l' Homme in Paris in September 2018. I would like to thank the Princeton University Center and the Maison des Sciences de l' Homme for hosting me as well as my intellectual home during the last years, the Department of History and Philosophy of Science in Athens.

It is my great pleasure to thank all those who have supported this project: above all Philip Kitcher for being an excellent interlocutor and for giving his consent that I use his works and ideas in this dialogue; Theo Arabatzis for very useful comments and moral support; Tassos Bougas and Stelios Virvidakis for their advice; Darrell Arnold for both his comments and his linguistic corrections; and Pablo Abitbol, Costas Dimitracopoulos, Uljana Feest, Gerd Gigerenzer, Anthony Hatzimoysis, Aristides Hatzis, Catherine Herfeld, Vasso Kindi, Eleni Manolakaki, Philip Pettit, Yannis Stefanou and Timothy Williamson for their encouragement and two anonymous reviewers for extremely useful comments. And I would like to apologize to all those who believe that writing a dialogue is neither a legitimate nor a fruitful way of philosophizing in our era.

I am indebted to my editor at Springer, Lucy Fleet, for her goodwill and support of the project. I would also like to thank Hans van Sintmaartensdijk at Springer for backing the project. I owe a special thanks to Arumugam Deivasigamani for his support during the production process.

I would like to thank the University of Minnesota Press for permission to publish excerpts from pp. 412, 415, 419, 424, 430–32, 437, 448 and 473 of the article by Philip Kitcher: "Explanatory Unification and the Causal Structure of the World", in: *Scientific Explanation, Minnesota Studies in the Philosophy of Science, Volume XIII*, edited by Philip Kitcher and Wesley Salmon, University of Minnesota Press, 1989.

Above all, I want to thank my family and especially my wife, Georgia, for her love and affection.

Athens, Greece   C. Mantzavinos
October 2018

Contents

Part I
On the Covering Law Model of Explanation

STUDENT: It is very nice to see you again, in Athens this time.

PHILIP: Yes, and I am so glad that you have suggested that we come to Cape Sounion. This is such a wonderful temple and the view over the Aegean is breathtaking.

STUDENT: Do you know how the Aegean Sea has got its name?

PHILIP: No, tell me.

STUDENT: Aegeus used to be the king of Athens. While visiting Athens, Androgeus, the son of King Minos of Crete, managed to defeat Aegeus in every contest during the Panathenaic Games. Out of jealousy, Aegeus sent him to conquer the Marathonian Bull, which killed him. Minos was angry and declared war on Athens. He offered the Athenians peace, however, under one condition: every nine years Athens would send seven young men and seven young women to Crete to be fed to a vicious monster, the Minotaur, who had a head of a bull on the body of a man. He dwelt at the center of the Cretan Labyrinth, an elaborate maze-liken construction designed by the architect Daedalus. On the third occasion Theseus, the son of Aegeus, volunteered to slay the monster and put an end to this horror for the Athenian youth. He took the place of one of the youths and set off with a black sail, promising to his father, Aegeus, that if he were successful he would return with a white sail. With the help of Ariadne, the daughter of King Minos of Crete, who fell in love with him, Theseus managed to kill the Minotaur. On the advice of Daedalus, Ariadne gave Theseus a ball of thread—a clue—and so he could find his way out of the Labyrinth. As the ship bearing Theseus and his liberated fellow Athenians approached the coast of Attica, Theseus forgot, such was the joy and excitement on the ship, to hoist the sail which was to have been the token of their safety to his father, Aegeus. King Aegeus watched daily for the return of his son—when he saw the ship's sail to be black threw himself from this rock here, from Cape Sounion, in despair. The sea became his name, Aegean Sea.

PHILIP: What a wonderful story! And what a beautiful explanation!

STUDENT: Indeed. The thing is, of course, that people used to believe in the truth of this explanation for quite a long time, probably for centuries!

© The Author(s) 2018
C. Mantzavinos, *A Dialogue on Explanation*, SpringerBriefs in Philosophy
https://doi.org/10.1007/978-3-030-05834-0_1

PHILIP: Mythical explanations are very charming, and I certainly prefer them to religious explanations which prevailed for an even longer time and which are certainly still with us.

STUDENT: I prefer scientific explanations to both mythical and religious ones, to be sure!

PHILIP: I certainly prefer those too, but everything hinges on what we mean when we call an explanation, a scientific one.

STUDENT: This is certainly not the time to discuss demarcation criteria of science vis-à-vis non-science, but still, tell me what do you think is the meaning of a scientific explanation, properly understood?

PHILIP: You know, I have thought about this for years and what I have come up with is that scientific explanation is very closely connected with unification.

STUDENT: Explanation as unification, then.

PHILIP: Exactly.

STUDENT: But isn't it the main message of Karl Popper's philosophy of science, that is, that explanations are arguments consisting of at least one scientific law and of initial conditions which together make up the explanans, and which necessarily lead to the conclusion, the explanandum, which is the sentence describing the phenomenon to be explained?

PHILIP: First of all, it was Carl Hempel[1] that formulated this model of explanation and not Karl Popper. Besides, this 'covering-law' model of explanation is probably the official model of logical empiricism, but there was also an unofficial model, a view of explanation which was not treated precisely, but which sometimes emerged in discussions of theoretical explanation. And this has to do with unification.

STUDENT: Sounds like 'Die ungeschriebene Lehre Platons'![2]

PHILIP: Not exactly, but probably similar.

STUDENT: Before telling me why unification is not part of the official doctrine of the logical positivists, let me ask you about the paternity of the model: wasn't it Popper who first introduced it?[3]

PHILIP: Be it as it may!

STUDENT: But now, tell me more about the unofficial model of logical empiricism!

PHILIP: Let me put it into a broader perspective and let's call it the 'Analytic Project'. The theory of explanation was only one of the three elements of this project.

STUDENT: The Vienna Circle claimed, of course, that there was a failure of philosophy and that the failure laid in the fact that philosophical debates focused on questions for which no resolution was in principle possible. Traditional philosophy was according to their diagnosis a mixture of pseudo-problems and real questions and the proper task of philosophy was to expose the pseudo-problems, and to liberate the serious issues in a way that would prepare for their solution.

PHILIP: The 'Analytic Project' is, indeed, a specification of what you have just said. The thought that the sciences obtain especially sound knowledge has inspired the search for a *theory of confirmation* that would delineate the general ways in which scientific hypotheses are supported by evidence. The thesis that scientists strive for clear and precise formulations of their ideas had sparked attempts to provide a *theory of theories and theoretical language.* Finally, the view that the sciences deepen our understanding of the natural world has given rise to proposals for a *theory of explanation.* These ventures constitute the 'Analytic Project'.

STUDENT: I agree, the theory of scientific explanation was part of a larger project, but it probably was the cornerstone of the 'Analytic Project'.

PHILIP: Yes, and it has certainly failed in the Popper-Hempel version of it. There are four major defects. One source of trouble lies in our propensity to accept certain kinds of historical narrative—both in the major branches of human history and in evolutionary studies—as explaining why certain phenomena obtain, even though we are unable to construct any *argument* proper that subsumes the phenomena under general laws. A second line of objection to the covering-law models is based on the difficulty in providing a satisfactory analysis of the notion of a scientific law.

STUDENT: Do you mean Goodman's paradox?[4]

PHILIP: Indeed. The challenge is to distinguish laws from mere accidental generalizations, not only by showing how to characterize the notion of a projectable predicate and thus answer Goodman's puzzle, but also by diagnosing the feature that renders pathological some statements containing only predicates that are intuitively projectable. For example 'No emerald has a mass greater than 1000 kg.'

STUDENT: Let me guess the third objection—it has got to do with the flagpole counter-example.

PHILIP: Exactly.

STUDENT: The so-called asymmetry problem. One main task of Hempel was to push causality as far as possible outside the discussion of scientific explanation.

PHILIP: He had good reasons to do this, and he was just standing in the long tradition of logical empiricists who wished to push out thorny metaphysical notions like that of causality from any discussions about science.

STUDENT: In the famous dictum of Bertrand Russell which I love: 'The law of causality, like much that passes muster among philosophers, is a relic of a bygone age, surviving, like the monarchy, only because it is erroneously supposed to do no harm.'[5]

PHILIP: Exactly.

STUDENT: But if you push metaphysics out of the front door, it will return from the back window.

PHILIP: This is what happened also in the theory of scientific explanation, as the counter-example that Sylvain Bromberger produced already in the 1960's can easily show.[6] We can explain the length of the shadow cast by a high object, a flagpole say, by deriving a statement identifying the length of the shadow from premises that include the height of the object, the elevation of the sun, and the laws of propagation of light. That derivation fits Hempel's model and appears to explain its conclusion. But, equally, we can derive the height of the object from the length of the shadow, the elevation of the sun and the laws of propagation of light, and the latter derivation intuitively fails to explain its conclusion. The challenge is to account for the asymmetry.

STUDENT: Besides asymmetry there is, of course, also the problem of the appeal to irrelevant factors.

PHILIP: Yes. If a magician casts a spell over a sample of table salt, thereby 'hexing' it, we can derive the statement that the salt dissolved on being placed in water from premises that include the apparently lawlike assertion that all hexed salt dissolves on being placed in water. But, it is suggested, the derivation does not explain why the salt dissolved.[7]

STUDENT: Exactly. And what is the fourth objection to the traditional Hempel-Popper account?

PHILIP: Do you know the cause of death of Friedrich Nietzsche?

STUDENT: He died eleven years after a breakdown brought on by paresis, I think.

PHILIP: Why did he have paresis?

STUDENT: He contracted syphilis from a prostitute, I think.

PHILIP: Indeed. Paresis, with its dreadful and lethal neurological decay, sometimes occurs when syphilis is untreated.

STUDENT: Is this the explanation of the tragic end of a great philosopher?

PHILIP: No. The account does not make Nietzsche's death probable.

STUDENT: Why not?

PHILIP: Nobody develops paresis without having contracted syphilis first, but the chances that untreated syphilis will lead to paresis are only about 15%. So, although the explanation makes the mode of death more probable, it does not show that it was to be expected, for the probability remains quite low.

STUDENT: And why is this problematic for the Popper-Hempel model?

PHILIP: Because the main idea in this model is encapsulated in the phrase of 'nomic expectability', and in the case of probabilistic explanations the requirement has been, of course, a requirement of high probability. Among the guiding ideas of this account of explanation was the proposal that explanation works by showing that the phenomenon to be explained was to be expected. In the context of statistical explanation of individual events, it was natural to formulate the idea by demanding that explanatory arguments confer high probability on their conclusions.[8]

STUDENT: But given that I can perfectly understand the explanation of Nietzsche's death, even if the probability is only about 15%, this model of explanation seems to be counter-intuitive.

PHILIP: Yes, this is the argument.

STUDENT: I can see that all these objections are lethal to the traditional account of explanation. My reason guides me to overthrow it, but it is such a pity!

PHILIP: Why?

STUDENT: Because it is so elegant and simple and does make sense in a series of cases, in which we try to figure out what constitutes the 'scientific' in a 'scientific explanation'.

PHILIP: But it is thrown into the Aegean Sea in the meanwhile.

STUDENT: What a shame! It was probably the last heroic attempt to build a philosophical account of scientific explanation by real heroes of the era!

PHILIP: Yes, this was a heroic attempt, but certainly not the last one. I think that the main shortcoming of the approach was that it did not categorize explanation as an activity.

STUDENT: This seems indeed correct.

PHILIP: In this activity we answer the actual or anticipated questions of an actual or anticipated audience. We do so by presenting reasons. We draw on the beliefs we hold, frequently using or adapting arguments furnished to us by the sciences.

STUDENT: I like the approach. Conceptualizing the whole issue in terms of activity seems to enable us to say more easily that explanation is just the other side of the coin of prediction and of manipulation, of actively changing the world.

PHILIP: But one has to be careful, of course. The relationship between explanation and practical control need not be that straightforward. The Copernican, heliocentric view of the solar system does not give us manipulable factors or practical control any more than the Ptolemaic system does. Explanations in geology—say, of the formation of continents—do not give us the ability to predict or control, but they are good explanations for all that.

STUDENT: I accept this point. But still, to view explanation as a kind of activity opens up a whole array of new issues.

PHILIP: It surely does!

STUDENT: I guess context becomes extremely relevant.

PHILIP: Exactly.

STUDENT: As in the following apocryphal story: when Willie Sutton was in prison, a priest who was trying to reform him asked him why he robbed banks. 'Well', Sutton replied, 'that's where the money is.'

PHILIP: Right! There is a failure to connect here, a failure of fit. Sutton and the priest are passing each other by. The challenge is to say how, exactly, they differ.

STUDENT: So, how do they differ exactly, then?

PHILIP: There are different values and purposes shaping the question and answer. They take different things to be *problematic* or to stand in need of explanation. The difference between them is that they have two different *contrasts* in mind, two different sets of alternatives to the problematic: Sutton robs banks. They are embedding the phenomenon to be explained in two different spaces of alternatives, which produces two different things-to-be-explained, two different objects of explanation.[9]

STUDENT: So, context is important! And explanatory relativity is introduced by way of contrastive explanations.

PHILIP: Yes, and the point of the example of Sutton shows the serious obstacles to any attempt to formulate a purely syntactical characterization of explanation.[10]

STUDENT: I see. The positivist separation of logic and pragmatics meant, in a sense, that «for many years pragmatics was the Cinderella of language, forced to stay home and do the dirty work while sisters syntax and semantics received all the attention».[11]

PHILIP: One can put it this way, yes! Van Fraassen has formulated the issue formally,[12] starting with the simple observation that why-questions are essentially contrastive: the question 'Why P?' is elliptical for 'Why P rather than P*, P**, …?'

STUDENT: This makes sense. So, we should pay attention to context? I guess that this is the main point.

PHILIP: Exactly. In the context in which the question posed, there is a certain body K of accepted background theory and factual information. This is a factor in the context, since it depends on who the questioner and audience are. It is this background which determines whether or not the question arises; hence a question may arise or conversely, be rightly rejected, in one context and not in another.

STUDENT: What van Fraassen suggests then, is that the discussion of explanation went wrong at the very beginning when explanation was conceived of as a relationship like description: a relation between theory and fact. Really it is a three-term relation, between theory, fact and context. No wonder that no single relation between theory and fact ever managed to fit more than a few examples!

PHILIP: Being an explanation is essentially relative on this account, for an explanation is an *answer*. In just that sense, being a daughter is something relative: every woman is a daughter, and every daughter is a woman, yet being a daughter is not the same as being a woman.

STUDENT: So to say that a given theory can be used to explain a certain fact is always elliptic for: there is a proposition which is a telling answer, relative to this theory, to the request for information about certain facts that bears on a comparison between this fact, which is the case, and certain contextually specified alternatives which are not the case.

PHILIP: In this picture scientific explanation is not pure science, but an application of science. It is a use of science to satisfy certain of our desires; and the desires are quite specific in a specific context, but they are always desires for descriptive information. Recall: every daughter is a woman.

STUDENT: You seem to be quite convinced by this account?

PHILIP: No, not quite. Although I think that van Fraassen has probably offered the best theory of the pragmatics of explanation, if his proposal is seen as a pragmatic theory of explanation, then it faces serious difficulties—most importantly that there is a lack of any constraints on 'relevance' relations which allows just about anything to count as the answer to just about any question.

STUDENT: I do not want to know about the technical details regarding the pragmatic theory of explanation and I am convinced that the official 'covering-law model' must be abandoned. Tell me at last more about your own account, the unofficial theory of logical positivism, tell me about explanation as unification.

PHILIP: Before I do that, let me say what I have been opposing when developing my unificationist account of explanation.

STUDENT: Who was your enemy?

PHILIP: He was not an enemy; he was a friend, Wes Salmon, who has given the great tradition of causal explanation a new lease of life.

STUDENT: *Amicus Plato, sed magis amica veritas.*

PHILIP: Yes, and since you are referring to Aristotle, he of course was the first to tie explanation to causation.

STUDENT: Standing on the holy ground of Attica, it is only natural to appeal to Aristotle!

PHILIP: Yes, Aristotle thought that causal knowledge is a superior type of knowledge, the type that characterizes science. What is characteristic of explanation is understanding the reason why, which is ultimately tied to finding the causes (αἴτια) of the phenomena. And one of the most famous distinctions in the history of philosophy is the one between four types of causes. The material cause is 'the constituent from which something comes to be'; the formal cause is 'the formula of its essence'; the efficient cause is 'the source of the first principle of change or rest'; and the final cause is 'that for the sake of which' something happens. For instance, the material cause of a statue is its material (e.g. bronze); its formal cause is its form or shape; its efficient cause is its maker; and its final cause is the purpose for which the statue was made.

STUDENT: Yes, in the famous passage of his Physics (194b23-195a3) and Metaphysics (Δ, 2):

Ἕνα μὲν οὖν τρόπον αἴτιον λέγεται τὸ ἐξ οὗ γίγνεταί τι ἐνυπάρχοντος, οἷον ὁ χαλκὸς τοῦ ἀνδριάντος καὶ ὁ ἄργυρος τῆς φιάλης καὶ τὰ τούτων γένη. Ἄλλον δὲ τὸ εἶδος καὶ τὸ παράδειγμα, τοῦτο δ' ἐστὶν ὁ λόγος ὁ τοῦ τί ἦν εἶναι καὶ τὰ τούτου γένη (οἷον τοῦ διὰ πασῶν τὰ δύο πρὸς ἕν, καὶ ὅλως ὁ ἀριθμός) καὶ τὰ μέρη τὰ ἐν τῷ λόγῳ. Ἔτι ὅθεν ἡ ἀρχὴ τῆς μεταβολῆς ἡ πρώτη ἢ τῆς ἠρεμήσεως, οἷον ὁ βουλεύσας αἴτιος, καὶ ὁ πατὴρ τοῦ τέκνου, καὶ ὅλως τὸ ποιοῦν τοῦ ποιουμένου καὶ τὸ μεταβάλλον τοῦ μεταβαλλομένου. Ἔτι ὡς τὸ τέλος· τοῦτο δ' ἐστὶν τὸ οὗ ἕνεκα, οἷον τοῦ περιπατεῖν ἡ ὑγίεια.

PHILIP: One modern manifestation of this long tradition of tying explanation to causation is Lewis's view that causal explanation of a singular event consists in providing some information about its causal history.[13]

STUDENT: Any particular event that we might wish to explain stands at the end of a long and complicated causal history. We might imagine a world where causal histories are short and simple; but in the world as we know it, the only question according to Lewis is whether they are infinite or merely enormous.

PHILIP: Yes, but in order to provide an adequate explanation you need only to carve a specific part of the causal net and provide it as an answer to the request of a why-question. An act of explaining may be more or less satisfactory, in several different ways according to Lewis, and his main point is that explanations are not things we may or may not have one of; rather, explanation is something we may have more or less of.

STUDENT: So, this is fundamentally different than Hempel who writes somewhere that '[t]o the extent that a statement of individual causation leaves the relevant conditions, and thus also the requisitive explanatory laws, indefinite it is like a note saying that there is a treasure hidden somewhere'.[14]

PHILIP: The note will help you find the treasure provided you go on working, but so long you have only the note you have no treasure at all; and if you find the treasure you will find it all at once. Lewis says it is not like that. "A shipwreck has spread the treasure over the bottom of the sea and you will never find it all. Every doublon you find is one more doublon in your pocket, and also it is a clue to where the next dublons may be. You may or may not want to look for them, depending on how many you have so far and on how much you want to be how rich".[15]

STUDENT: But Salmon's account does not stress causal history so much, does it?

PHILIP: There is certainly a historical element in it, but the main idea is that of a causal mechanism, of course.

STUDENT: His distinction between three approaches to scientific explanation, all going back to Aristotle, has always interested me. The *epistemic conception* is essentially the Popper-Hempel approach since it makes the concept of explanation broadly epistemic, taking explanation to be nomic expectability. The *modal conception* has an account of necessity as its centerpiece: the explanandum is said to follow necessarily from the explanans—things could not have happened otherwise given the relevant laws. According to the *ontic conception* that Salmon propagates, to give scientific explanations is to show how events fit into the causal structure of the world.[16]

PHILIP: Yes, it is the ontic conception that he prefers, distancing himself from van Fraassen's pragmatic approach by means of a simple observation: it does not follow that once we have ascertained by reference to pragmatic and contextual factors what explanation is being sought, that the explanation itself must embody pragmatic features.

STUDENT: Does he give a concrete example?

PHILIP: Yes. His example is the following: "Suppose that a member of a congressional committee asks why a particular airplane that took off from Washington National Airport one winter day failed to gain sufficient altitude, with the result that it struck a bridge and crashed into the Potomac River. An official investigator answers that it was because ice had built up on the wing surfaces. The questioner is satisfied. The investigator goes on, nevertheless, to explain that, due to the Bernoulli principle, an airplane derives its lift from wings that have the shape of airfoils. Ice adhering to the leading edge of the wing alters the shape of the surface so that it no longer functions as an effective airfoil, and no longer provides adequate lift. The committee member, uninterested in scientific details, dismisses the latter part of the explanation as irrelevant—in van Fraassen's terms, it is not salient to this person in this context. Hence, van Fraassen would maintain that the information about the Bernoulli principle, the shape of the airfoil and the lift does not constitute an acceptable part of the answer to the why-question posed by the member of Congress, even though that information does not constitute part of that person's background knowledge.

The fact that the airplane went down in the river is not explained [on Salmon's view] unless the causal mechanisms relating the accumulation of ice on the wings to the loss of lift are mentioned. For van Fraassen, in contrast, the buildup of ice on the wings is an adequate explanation if that information fills the gap in the questioner's knowledge that he or she wanted to have filled. The scientific elaboration furnishes explanatory knowledge only if the questioner has scientific curiosity; otherwise it is not salient. This is the point at which van Fraassen's pragmatic view is untenable".[17]

STUDENT: And the novelty of Salmon's approach to causality consists in taking processes rather than events as basic entities, I gather.

PHILIP: Yes. The main difference between events and processes is that events are relatively localized in space and time, while processes have much greater temporal duration, and in many cases, much greater spatial extent.

STUDENT: And what is a causal mechanism?

PHILIP: Salmon offers a generic account of causal mechanism, based on two key central concepts, causal process and causal interaction. Causal processes are the mechanisms that propagate structure and transmit causal influence in this dynamic and changing world. They provide the ties among the various spatiotemporal parts of the universe.[18]

STUDENT: Give me a couple of examples.

PHILIP: The movement of a ball or a light-wave travelling from the sun.

STUDENT: What about causal interaction?

PHILIP: Before telling you about causal interaction, let me add that processes are continuous, so that a process cannot be represented as a sequence of discrete evens. The continuity of the process provides the direct link between cause and effect.

STUDENT: I see.

PHILIP: Not all processes are causal, but only those that are capable of transmitting a mark. A mark is a modification of the structure of a process. And a process is causal, if it is capable of transmitting the modification of its structure that occurs in a single local interaction.

STUDENT: But isn't the transmission of a mark a bit mysterious?

PHILIP: No, there is no mystery involved in the view that a mark is transmitted from a point A of a process to a subsequent part B, if we invoke Russell's 'at-at' theory of motion.

STUDENT: You mean the one that Russell developed as a reply to Zeno's paradox of the arrow?

PHILIP: Exactly.

STUDENT: I think the main claim of the 'at-at' theory is that to move from A to B is simply to occupy the intervening points at the intervening instants. It consists in being *at* particular points of space *at* corresponding moments. There is no *additional* question as to how the arrow *gets from* point A *to* point B; the answer has already been given—by being at the intervening points at the intervening moments.

PHILIP: When two processes intersect, and they undergo correlated modifications that persist after the intersection, Salmon says that the intersection constitutes a causal interaction. This is the basic idea behind what he wants to take as a fundamental concept.[19]

STUDENT: Is, then, mark transmission necessary and sufficient for a process being causal?

PHILIP: It is neither. Take the case of a pseudo-process like the shadow of a moving car. This can be permanently marked by a single local interaction: when the car grazes the wall its shadow acquires, and continues to have, a new characteristic—being the shadow of a scratched car. So, the mark-transmission is not sufficient for a process being causal. Conversely, a process can be causal, even if there is no mark-transmission involved. Consider Salmon's requirement that a process should remain uniform with respect to one characteristic for some time. This requirement, however, seems to exclude from being causal many genuine processes that are short-listed like the generation and annihilation of virtual subatomic particles.

STUDENT: Can this theory be saved?

PHILIP: Dowe has tried to save it by suggesting that it is the possession of a conserved quantity, rather than the ability to transmit a mark, that makes a process a causal process.[20]

STUDENT: But even if we grant that this offers a plausible account of physical causal mechanisms, it can be defended as a general theory of causal mechanisms only if married to a strong reductionism that all worldly phenomena, be they biological, psychological or social, are reducible to physical phenomena.

Part II
On Explanation as Unification and Other Models of Explanation

PHILIP: Listen, the invocation of causal notions has its costs. Hempel's account of explanation was to be part of an empiricist philosophy of science and it could therefore only draw on those concepts that are acceptable to empiricists. If causal concepts are not permissible as primitives in empiricist analyses, then either they must be given reductions to empiricist concepts or they must be avoided by empiricists. Hempel's work appears to stand in a distinguished tradition of thinking about explanation and causation, according to which causal notions are to be understood in terms of concepts that are themselves sufficient for analyzing explanations. Empiricist concerns about the evidence that is available for certain kinds of propositions are frequently translated into claims about conceptual priority. Thus, the thesis that we can only gain evidence for causal judgments by identifying lawlike regularities generates the claim that the concept of law is prior to that of cause, with consequent dismissal of analyses that seek to ground the notion of law in that of cause.

One of Hume's legacies[21] is that causal judgments are epistemologically problematic. For those who inherit Hume's theses about causation (either his positive or his negative views) there are obvious attractions in seeking an account of explanation that does not take any causal concepts for granted. A successful analysis of explanation might be used directly to offer an analysis of causation—most simply, by proposing that one event is causally dependent on another just in case there is an explanation of the former that includes a description of the latter. Alternatively, it might be suggested that the primitive concepts employed in providing an analysis of explanation are just those that should figure in an adequate account of causation.

Because the invocation of causal dependency is so obvious a response to the problems of asymmetry and irrelevance, it is useful to make explicit the kinds of considerations that made that response appear unavailable. My main claim is that there is a tension between two attractive options. Either we can have a straightforward

© The Author(s) 2018
C. Mantzavinos, *A Dialogue on Explanation*, SpringerBriefs in Philosophy
https://doi.org/10.1007/978-3-030-05834-0_2

resolution of asymmetry problems, at the cost of coming to terms with epistemological problems that are central to the empiricist tradition, or we can honor the
constraints that arise from empiricist worries about causation and struggle to find
some alternative solutions to the asymmetries.

STUDENT: So, the central issues that arise concern the adequacy of proposed
epistemological accounts of causation and of suggestions for overcoming problems
of asymmetry and irrelevance without appealing to causal concepts.

PHILIP: Exactly.

STUDENT: And so you can tell me at last more about explanation as *unification*!

PHILIP: Yes, but let us find a more shady spot first.

STUDENT: Let us walk down this path, then.

PHILIP: Here we are. The shadow of this tree is very pleasant—and it is very
cool here. So, the crucial point is that the 'because' of causation is always derivative
from the 'because' of explanation.

STUDENT: And this seems to imply that there are explanations which are not
causal, right?

PHILIP: Exactly.

STUDENT: Give me an example!

PHILIP: Mathematical explanation! You should give me a pen and a piece of
paper, because it is getting a bit more serious now and I need to show you how this
works!

STUDENT: Here they are.

PHILIP: Let me give a concrete example, so that you can see the point. It is
possible to axiomatize the theory of finite groups in a number of different ways.
The standard approach takes a group to be a finite set which is closed under an
associative operation, multiplication. There is an independent element, 1, such that
for any element α of the group $\alpha 1 = 1\alpha = \alpha$. And for any element α there is an
inverse α^{-1} such that $\alpha^{-1}\alpha = \alpha\alpha^{-1} = 1$. On this basis one can prove that division
is unique wherever it is possible. Alternatively, a finite group can be identified as a
finite set, closed under an associative operation, multiplication, such that division is
unique wherever it is possible. On this basis, one can show that finite groups have
all the properties attributed in the usual axioms. Mathematicians distinguish the two
axiomatizations—if only in their practice of choosing the standard axioms—and they
sometimes express the preference by saying that the usual axioms are more 'natural'
than the nonstandard ones, or that the division property is 'less fundamental'. I
suggest that what they recognize is a case of the asymmetry problem: we can explain
why finite groups satisfy the division property by using the axioms about the existence
of inverses and idempotent elements to demonstrate that division is unique wherever
it is possible. But the derivation of the existence of an idempotent element and of
inverses from the division property is nonexplanatory, and, I think, nonexplanatory in
just the same way as the derivation of heights from shadow lengths. If this is correct,
then the example reveals that the asymmetry problem can arise in cases where causal
considerations are quite beside the point.

STUDENT: This is a wonderful example! I fully accept the point: not all explanation is causal.[22]

PHILIP: I could give you many more examples, but since you are already convinced, let me now elaborate a bit more on explanation as unification. The heart of my view is that successful explanations earn that title because they belong to a set of explanations, the explanatory store, and that the fundamental task of a theory of explanation is to specify the conditions on the explanatory store. Intuitively, the *explanatory store* associated with science at a particular time contains those derivations which collectively provide the best systematization of our beliefs. Science supplies us with explanations whose worth cannot be appreciated by considering them one-by-one, but only by seeing how they form part of a systematic picture of the order of nature.

STUDENT: Can you make it more precise?

PHILIP: First of all I accept that explanations are *arguments*, as Hempel suggested. But the conception of argument as a *premise-conclusion* pair is too narrow. On the systematization account, an argument is considered as a derivation, as a sequence of statements whose status is clearly specified.

STUDENT: This seems plausible.

PHILIP: For a derivation to account as an *acceptable* ideal explanation of its conclusion in a context where the set of statements endorsed by the scientific community is K, that derivation must belong to the explanatory store over K, $E(K)$. So, $E(K)$ is to be the set of derivations that best systematizes K, and I suggest that the criterion for systematization is unification. $E(K)$, then, is the set of derivations that best unifies K.

STUDENT: The challenge, of course, is to say as precisely as possible what this means.

PHILIP: Here is the main message: Science advances our understanding of nature by showing us how to derive descriptions of many phenomena, using the same patterns of derivation again and again, and, in demonstrating this, it teaches us how to reduce the number of types of facts we have to accept as ultimate (or brute).

STUDENT: Although I like your main message, I dislike the part of it referring to the advancement of our understanding. This is similar to what Michael Friedman has once proposed[23] and to a view many people seem to share today[24]; but this introduces a psychological concept in what is or should be an objective theory of explanation.

PHILIP: Well, I don't see what is so wrong with introducing a psychological notion into the account. I am certainly not a defender of an apriori notion of scientific explanation!

STUDENT: The problem is not with the apriori character of a theory of explanation, but rather with the subjective element that you are introducing here. If the aim of science is merely to advance our *understanding* of nature, then what guarantees that we ever reach a correct account of nature? I can understand nature by imposing deities behind all natural phenomena, but this would surely not provide me a correct account of those phenomena!

PHILIP: But I did not say that science would advance our understanding of nature by *any* means, but by very specific means: by showing us how to derive descriptions of many phenomena, using the same patterns of derivation again and again. I cannot see how imposing deities would allow any such derivations to be formulated!

STUDENT: I accept that. I have, however, a general distaste of the employment of the notion of understanding in discussions about scientific explanation, because this involves the risk of falling back to the old discussion of the dualism between *Verstehen und Erklären*.

PHILIP: But this need not be the case. And if you let me proceed with a formal characterization of this general message, you will see that explanation as unification, even if it employs the notion of (scientific!) understanding, can be perfectly objective!

STUDENT: I am eager to hear!

PHILIP: First we need the notion of pattern. A *schematic sentence* is an expression obtained by replacing some, but not necessarily all, the nonlogical expressions occurring in a sentence with dummy letters. Thus, starting with the sentence 'Organisms homozygous for the sickling allele develop sickle-cell anemia', we can generate a number of schematic sentences: for example 'Organisms homozygous for A develop P' and 'For all x, if x is O and A then x is P' (the last being the kind of pattern of interest to logicians, in which all the nonlogical vocabulary gives way to dummy letters). A set of *filling instructions* for a schematic sentence is a set of directions for replacing the dummy letters of the schematic sentence, such that, for each dummy letter, there is a direction that tells us how it should be replaced. For the schematic sentence 'Organisms homozygous for A develop P', the filling instruction might specify that A be replaced by the name of an allele and P by the name of a phenotypic trait. A *schematic argument* is a sequence of schematic sentences. A *classification* for a schematic argument is a set of statements describing the inferential characteristics of the schematic argument: it tells us which terms of the sequence are to be regarded as premises, which are inferred from which, what rules of inference are used, and so forth. Finally, a *general argument pattern* is a triple consisting of a schematic argument, a set of sets of filling instructions, one for each term of the schematic argument, and a classification for the schematic argument.

STUDENT: And now it's time for a concrete example, I think.

PHILIP: Yes, but as a prelude let me recall an important point made by Thomas Kuhn. When we conceive of scientific theories as sets of statements (preferably finitely axiomatized) then we naturally think of knowing a scientific theory as knowing the statements—typically knowing the axioms and, perhaps, some important theorems. But, as Kuhn points out, even in those instances where there are prominent statements that can be identified as the core of the theory, statements that are displayed in texts and accompanied with names—as for example, Maxwell's equations, Newton's laws, or Schrödiger's equation—it is all too common for students to know the statements and yet fail to understand the theory.[25]

STUDENT: Well, it is true that I often could not do the exercises at the end of the chapter!

PHILIP: This is because scientific knowledge involves more than knowing the statements. A good account of scientific theories should be able to say what the extra cognitive ingredient is.

STUDENT: So what is this extra cognitive ingredient, then?

PHILIP: I claim that to know a theory involves the internalization of the argument pattern associated with it, and that, in consequence an adequate philosophical reconstruction of a scientific theory requires us to identify a set of argument patterns as one component of the theory. This is especially obvious when the theory under reconstruction is not associated with any 'grand equations', like the Darwinian evolutionary theory, my favorite example.

STUDENT: But Darwin's theory of evolution by natural selection addresses a number of general questions about human life[26]—not just one.

PHILIP: Exactly. These questions include problems of biogeography, of the relationships among organisms (past and present), and of the prevalence of characteristics in species or in higher taxa. Dawin's principal achievement consisted in his bringing these questions within the scope of biology, by showing, in outline, how they might be answered in a unified way.

STUDENT: Explanation as Unification, ein schönes Gedankengebäude!

PHILIP: And I hope that you find illuminating that the explanatory store contains only deductive arguments, so that in a certain sense all explanation is deductive. I call this deductive chauvinism!

STUDENT: Quite an aggressive naming strategy! But the crucial point can be stated with sobriety: the 'because' of causation is always derivative from the 'because' of explanation.

PHILIP: Causalists, most prominently Salmon among them, claim that explanation consists in identifying causal relations. This is a 'bottom-up' approach. Causal relations primarily relate individual events or, if one prefers a process ontology, the depiction of causal processes is the fundamental issue in the provision of explanation. My approach (and Hempel's) is 'top-down'. Explanatory concepts are conceived as prior to causal concepts. The starting point is the idea that explanation is directed at an ideal of scientific understanding. We achieve that ideal by giving a unified, deductive systematization of our beliefs. Our views about genuine properties and explanatory dependence emerge from the project of unifying the regularities we discover in nature. On this approach, theoretical explanation is primary. Causal concepts are derivative from explanatory concepts. In explaining particular events we answer as many questions as we can, drawing on our view of the order of natural phenomena. In some cases, an ideal of understanding may not be completely realizable.

STUDENT: Explanation as unification is an ingenious alternative to the traditional causal approach and to Salmon's causal-mechanical approach. But what about other approaches, like the mechanistic approach of Machamer et al. for example?

PHILIP: But this does not add anything essentially new to Salmon's approach. Besides, it seems to be a very weak conception overall.

STUDENT: Why? Their definition seems important: 'Mechanisms are entities and activities organized such that they are productive of regular changes from start or set-up to finish or termination conditions'?[27]

PHILIP: But this is very shallow!

STUDENT: Is this supposed to be an argument?

PHILIP: I leave it to you to decide.

STUDENT: There are many defenders of this or a similar approach in diverse disciplines: in biology, in cognitive science, in the social sciences...

PHILIP: But not many in physics, not many in physics!

STUDENT: And what about Woodward's approach to explanation? Although this is broadly in the tradition of causal explanation it is quite distinctive, I think.[28]

PHILIP: Yes, it is a manipulationist account of causal explanation based on two key ideas, that of intervention and that of invariance. Intervention is about the following two characteristics: (a) the change of the value of X is entirely due to the intervention I and (b) the intervention changes the value of Y, if at all, only through changing the value of X. The notion of invariance is closely related to the notion of intervention. A generalization G relating, say, changes in the value of X to changes in the value of Y, is invariant if G would continue to hold under some intervention that changes the value of X in such a way that, according to G, the value of X would change—'continue to hold' in the sense that G correctly describes how the value of Y would change under this intervention. A necessary and sufficient condition for a generalization to describe a causal relationship is that it be invariant under some appropriate set of interventions.

STUDENT: I guess that the notion of invariance under interventions is intended to do the work that is done by the notion of a law of nature in other philosophical accounts.

PHILIP: Indeed.

STUDENT: On this account F = mg would count as a generalization and not as a candidate of a law of nature.

PHILIP: Yes, since it lacks many of the features standardly ascribed to laws: it holds only approximately, even near the surface of the earth, and fails to hold even approximately at sufficiently large distances from the earth's surface. And it is obviously contingent on the earth's having a particular mass and radius.

STUDENT: So, the traditional criteria of lawfulness need not be satisfied: exceptionlessness, necessity, breadth of scope or degree of theoretical integration.

PHILIP: This is in large measure a consequence of the ways in which invariance is defined: for a generalization to be invariant, all that is required is that it be stable under *some* range of changes and interventions. It is not required that it be invariant under *all* possible changes and interventions.

STUDENT: I see. One important element of the manipulationist approach is that causal claims relate variables. This seems straightforward. Another is that the notion of invariance is obviously a modal or counterfactual notion, since it has to do with whether a relationship would remain stable if, perhaps contrary to actual fact, certain changes or interventions were to occur.

PHILIP: Exactly. And this is the most problematic part of the theory. We know, of course, all the familiar attempts to provide a semantics for counterfactual conditionals, Lewis's being probably the most well-known one, that appeal to possible worlds.

STUDENT: But Woodward is very knowledgeable about actual scientific practices to accept such a view!

PHILIP: He is indeed very careful to avoid the metaphysical excesses of Lewis's theory[29] and also very careful in his overall use of counterfactuals. For him, only counterfactuals that are related to interventions can be of help. Those counterfactuals of a very special sort have to do with the outcomes of hypothetical interventions.

STUDENT: But still something must be said about their truth-conditions. What is it that makes a counterfactual true, if one does not want to invoke the notion of possible worlds?

PHILIP: This is an old problem. The situation with respect to counterfactuals and modality generally is distressingly similar to a predicament in the philosophy of mathematics that Benacerraf presents in compelling fashion[30]: our best semantic accounts and our best epistemological views do not cohere. For the best semantic accounts make reference to possible worlds, our best epistemological views make knowledge (and justification) dependent on the presence of natural processes that reliably regulate belief, and it is to say the least unobvious, how any natural process could reliably regulate our beliefs about possible worlds.

STUDENT: This is an extremely valuable point that you are making. But, still, I do not think that Woodward would like to tie the counterfactuals to any kind of possible worlds speculations. Consider a case from medical practice: a large group of people suffering from a disease are randomly divided into a treatment and a control group, with the former receiving some drug that is withheld from the latter, and the right sort of experimental controls being followed. If this experiment were actually carried out and the incidence of recovery was much higher in the treatment group than in the control group, it would be natural to think of it as providing good evidence for the truth of counterfactuals like the following: 'If those in the control group had received the drug, the incidence of recovery in that group would have been much higher.'[31]

PHILIP: I see, in this case the counterfactual could be true or false depending on the results of an actual experiment.

STUDENT: This is the point, I think, and it is well-taken. So, Woodward's approach has its merits in certain cases.

PHILIP: As does Streven's kairetic account.

STUDENT: This sounds like it is inspired by the ancient Greek καιρός, meaning a crucial moment.

PHILIP: A causal model is an explanation in case only if the model is not missing parts, and second, when every aspect of the causal story represented by the model makes a difference to the causal production of the explanandum. When a causal model that is not self-contained, or that mentions non-difference-makers, is presented as an explanation, then a false claim is made.

STUDENT: This sounds very similar to Mill[32] and Mackie.[33]

PHILIP: Yes, but it is substantially tied to the notion of *Depth*[34]: an explanation is deep when it drills down to the explanatory foundational level, which is the web of relations of causal influence orchestrated by the fundamental physical laws. So, physical depth lies at the center of the kairetic account, claiming that no explanation can stand alone unless it articulates the properties of the fundamental physical relations in virtue of which the phenomenon to be explained occurs.

STUDENT: So, this is nothing else than a reductionist account of explanation—very weak indeed, especially if it is also supposed to provide a model for the social sciences. Consider an example from economics. The macro-phenomenon of inflation, i.e. of the average growth of the level of prices in an economy over the years, is normally (and convincingly) explained by the amount of the quantity of money in the economy, assuming a constant money velocity. If there is a generally accepted explanation in economics at all, then this is the one. Now, according to the kairetic account, this should not count as an explanation at all, because it is not 'deep' enough, that is, it does not make any appeal to the causal influence of physical laws. And this is absurd.

PHILIP: Well, you have to make a distinction here which is valuable: the point is that we just do not currently avail of an explanation which employs physical laws, but this is not *in principle* impossible. It is only our current epistemological predicament: we do not know such an explanation yet, but we might avail of it in the near or distant future.

STUDENT: This is a general point about all our knowledge, not only our explanatory knowledge!

PHILIP: An important point about many social science generalizations is that they talk about things that are *social* facts—true because people agree they are. Now any reduction of such generalizations would not be a reduction of *items* in the generalizations to their physical make up; it would be a reduction of many of their properties to the *mental states* of people interacting with the items. Such a reduction might or might not work—but it should be noted that that's how the reduction would go. It's not primarily about reduction of stuff to physical stuff!

STUDENT: Be that as it may. What concerns me more is this passion to view all phenomena in a vertical way, top-down, rather than in a horizontal way, side-to-side. Why should we always search deeper in order to be satisfied, rather than look at one phenomenon from different perspectives?

PHILIP: I agree.

STUDENT: Besides, causal fundamentalism can certainly go hand in hand with explanatory ecumenism, as I once heard Philip Pettit and Frank Jackson[35] argue. It has been taken for granted, they think, both in the natural and in the social sciences, that those explanations should be judged better that provide fine-grained information and therefore greater detail in the relevant causal structure. In other terminology, there has been an *opinio communis* that an explanation of a macro-phenomenon should only be judged as successful if a causal structure at the micro-level is clearly identifiable. More information about the micro-level will supposedly make any explanation at the macro-level more acceptable. The search for micro-foundations is concomitant with a traditional, hierarchical view of scientific knowledge, with physics lying at the deepest level and the social sciences on the top—the search for micro-foundations goes hand in hand with a reductionist strategy meant to shed better and brighter light on the phenomenon in question.

PHILIP: If one takes the example that you referred to previously, the imperative to search for micro-foundations in order to provide a more compelling explanation of the phenomenon of inflation would require the provision of information on the behavior of households and firms in the economy—only these kinds of explanations would count as satisfactory and, thus, scientifically acceptable.

STUDENT: Exactly. And here is their crucial distinction between causal fundamentalism and explanatory fundamentalism. It is true that going micro and getting at smaller levels of causal grain involves getting better and better contrastive information—greater and greater detail—on causal history. But it does not follow that it involves getting better and better information *tout court*. On the contrary, the obvious thing to say is that while it means getting better and better contrastive information, it means losing information of a comparative kind. Thus, there is no reason to think that finding smaller and smaller levels of causal grain means better and better explanations. There is no argument that we should not be concerned as much with comparative as with contrastive information in our exploration of causal structure.

PHILIP: To return to our example, the rise of the money supply leading to the increase of inflation would count as comparative information; and this would be judged as equally important to the contrastive information provided by the description of the behavior of households and firms which leads to the emergence of the same phenomena.

STUDENT: The thrust of the argument of Pettit and Jackson has been that explanations at different levels, like explanations at different removes in time, may provide different sorts of information on the causal process that leads to the event that is to be explained. Thus, we should be open to the possibility that they each are interesting on their own terms. We may be committed to causal fundamentalism, i.e. the view that, in order to be acceptable, an explanation must offer a diagnosis of the causal structure, but we can at the same time endorse a pluralism on explanatory matters. I think that this was the first clear articulation of the position of explanatory pluralism or explanatory ecumenism in their terminology.

PHILIP: I certainly endorse explanatory pluralism.

STUDENT: I find this surprising! The whole time we have been sitting under the shadow of this poplar-tree, you have been talking about the merits of explanation as unification!

PHILIP: Yes, but taking stock of our discussion up to now, it seems that the search for '*the* structure of explanations' has not delivered a general account.

STUDENT: *D'accord*. But I must confess that an explanatory ecumenism tied to a causal fundamentalism does not seem to be very liberal in the end. A genuinely pluralistic position can and should make as few commitments as possible to causality and other metaphysical issues. Besides, explanatory pluralism need not hinge on positions akin to reductionism. The passion for viewing scientific phenomena through the lenses of reductionism is a relic of logical positivism and of the epoch of the explanatory monarchy of physics—which designated all other sciences as 'special'[36]—rather than of the explanatory liberalism of our times!

PHILIP: This is a quite strong thesis!

STUDENT: It is. But after all we have been discussing up to now, it seems that articulating explanatory pluralism in concreto is the main desideratum.

PHILIP: But we can't throw all these models of scientific explanation away! There is no reason for doing so! They seem to work in many cases.

STUDENT: Yes, they certainly do. But they are all unitary models of explanation. Their resources cannot capture anything but the explanatory activities of *some* areas of theoretical science. The claim that each of them raises, i.e. that it is supposed to accommodate *all* and *every* scientific activity, is not tenable.

PHILIP: I would certainly agree with that with respect to explanation as unification, but this must be shown first, before one proceeds to the adoption of a pluralistic stance.

STUDENT: Indeed. Let me give you three examples of current social scientific practice in order to convince you. Take neoclassical microeconomic theory which avails of a well-developed mathematical formulation and has a great range of applications. The general theoretical framework is that of the rationality hypothesis which is used to cover a great array of phenomena. The hypothesis of utility maximization plays the fundamental role in explaining the case of exchange of goods and services between two or more individuals under conditions of scarcity. The general argument pattern that lies at the heart of the neoclassical microeconomic theory is one of maximization under constraints. But there is a long series of specific argument patterns: the utility maximization of consumers, the profit maximization of firms, the vote maximization of politicians in democratic politics…

PHILIP: More and more specific argument patterns can be embedded into the larger argument pattern of maximization under constraints and thus provide explanations by means of unification.

STUDENT: Yes, and it is clear that only your unification theory of explanation can accommodate the practices of these scientists. Alternative models cannot. Microeconomics does not pay attention to causal processes, causal interactions or causal mechanisms, and there is no causal talk included.

PHILIP: It is not?

STUDENT: No. The endeavor is to determine precisely the equilibrium price and quantity without paying any attention to causal interactions. And the well-worked out deductive arguments employed to analyze market behavior are not amenable to an analysis of the manipulationist approach. The equations of neoclassical microeconomic theory are not invariant generalizations as the manipulationist model of explanation would require.

PHILIP: So, this is yet another type of deductive chauvinism!

STUDENT: Indeed! But of course there are other categories of social scientific practice where the causal model seems appropriate *and* the other models inappropriate.

PHILIP: Such as?

STUDENT: Such as the attempts of sociologists and political scientists to identify social mechanisms of only local validity. For example, the well-observed phenomenon in the liberal democracies of the Western world that the expansion of
educational opportunity does not increase social mobility nor reduce social equality.
Raymond Boudon has identified a widely accepted causal mechanism which figures
out the choices that the individual in each educational system has to make about
staying or not staying at school.[37] There is a primary effect that directly influences
the educational performance of the children as a joint result of class-specific primary
socialization processes such as the potential of parents to actively support their child
in school, the general cultural differences, etc. The other causal mechanism called
the secondary effect runs as follows: the prospects of success of children of working
class families in higher education tracks are estimated as rather low by their parents since the parents are mostly not familiar with this type of school themselves.
A major interest of the parents themselves which influences decisively their decision making is status maintenance. They want to make sure that their current status
remains stable. A consequence of this is that their children receive sufficient praise
and acceptance in their family environment even if they do not necessarily aim at
the highest academic achievement. Parents in families of a higher social status on
the other hand, are motivated to provide their own children with the best possible
education in order to make sure that their intergenerational status level remains relatively stable. These two causal mechanisms, the primary and the secondary effect,
result in a situation in which, because of family socialization, the children of parents
of unequal economic and educational background end up taking unequal advantage
of the existing educational opportunities. This mechanism gives rise to parents from
unequal social status producing children through a family socialization process that
make use of unequal educational opportunities. In a nutshell, even though education
does provide opportunity for individuals to improve their social position, this does
not translate into change in social structure.

PHILIP: I see, the understanding that the identification of such causal mechanisms
allows is 'local' rather than 'global' in nature.

STUDENT: Exactly. Raymond Boudon's explanation does not apply to societies
outside the Western liberal democracies and does not aspire to provide insights into
the causal structure of the educational systems of, say, the developing countries
or those of the Middle Ages. And there is no formalized argument pattern, nor is
it embedded in a more general one. So, the unificationist model of explanation is
not applicable here. Nor is the manipulationist model, since neither the notion of
intervention is employed or presupposed nor is any type of counterfactual reasoning.

PHILIP: Wait a minute! But what explains things here is just a universal rational choice model (or some other universal psychological principles) but applied to
particular circumstances—as ALL scientific laws are!

STUDENT: No it doesn't! At least the second causal mechanism that Boudon
is stating is not universal in nature! But let me give you a third and last example
which just occurred to me: econometrics. This is an established branch of quantita-
tive research in which functional relationships between dependent and independent

variables are sought and established. The unification model is not applicable here. Nor is the causal mechanical model, since the generalizations offered in econometrics, though they do invoke causal talk and causal inference, they never identify a causal mechanism proper. It is the manipulationist notion which is the only account which can accommodate this part of scientific practice.

PHILIP: The upshot of the discussion, I take it, is that the three main models of scientific explanation currently available, the unificationist, the causal-mechanical and the manipulationist can only accommodate some of the existing scientific practices.

STUDENT: Exactly. None of them can successfully claim the monopoly as the one and correct account of scientific explanation. These are just unitary models!

PHILIP: Is this the reason for their failure, you think?

STUDENT: No. I think the reason for their failure lies deeper—in the attempt to provide an answer to a wrong question: 'What is an explanation?'

PHILIP: But why is it a wrong question? This is the question that philosophers have been concerned with for centuries.

STUDENT: Because we can never provide the necessary and sufficient conditions of the concept of scientific explanation. Fixing its meaning might be somehow valuable, no doubt, but the core of the philosophical enterprise is to solve the descriptive and normative problems of scientific activity.

PHILIP: And what is the purpose of such a philosophy?

STUDENT: The goal of a philosophical account of explanation should not be to capture *the* explanatory relation, but rather to capture the many ways in which explanations are provided in the different domains of science—and of everyday life.

PHILIP: But there are serious philosophical dangers which may destroy such a pluralistic account!

STUDENT: Dangers are welcome. Surpassing them makes us stronger.

PHILIP: If such an alternative philosophical account exhausts itself in discovering reasons in order to legitimate the plurality of the explanatory practices, this will be nothing else than an apologetic exercise!

STUDENT: No apologies intended! Developing normative standards is always part of the philosopher's game!

PHILIP: Another danger is that this might christen as acceptable, any explanation offered in the scientific discourse by giving up any kind of scientific rationality whatsoever!

STUDENT: I am not sure that this is a distinct kind of danger, but I am certainly not an anarchist! All that I am suggesting is a genuine middle-ground position. For a very long time philosophers have constructed ideals far removed from the practices of the scientists. Those practices were supposed to measure up to these ideals—otherwise, they were to be discarded. Then historians, sociologists and weak spirited charlatans[38] entered the scene, maintaining that all that we should take seriously are those very practices: recording and accepting them were supposedly the philosophical tasks.

PHILIP: Epistemology without history is blind.

STUDENT: Yes. And the goal is now to find the middle way. At every moment of time there is a stock of explanations available in every society that are proposed by people organized formally in organizational structures, like churches, universities, etc. But explanations are also proposed by ordinary people in their everyday lives. The terms of their provision, criticism and dissemination are regulated by the *rules of the specific explanatory game.*

PHILIP: Explanatory game?

Part III
On Explanatory Games

STUDENT: Explanations emerge as participants play a specific explanatory game following the respective rules of the game. Different rules make up different games and they constrain the explanatory strategies that the participants are allowed to undertake.

PHILIP: But the rules cannot merely *constrain* the participants, I guess. At the end they are *enabling* them to provide the requested explanations.

STUDENT: Rules should be also seen along this enabling dimension—this is certainly important, since they simultaneously embed the entire history of the game.

PHILIP: How exactly?

STUDENT: Explanation should be seen as a process unfolding in historical time rather than as an outcome. The task is to highlight the complex process of explanation rather than to pose the issue as if explanation were static. During an evolutionary process of trial and error, explanatory activities are undertaken according to the rules of the game and a prevalent flow of explanations is produced, tested and retained or discarded. Novelty is a permanent feature of the process, as is the collective learning of the participants who adopt those rules that help them provide solutions to their respective explanatory problems.

PHILIP: Yes, but what does the historicity of the whole enterprise refer to?

STUDENT: To the learning history of the individual members of the respective group regarding which explanatory strategies are allowed and which are not. The best way to analyze the slogan 'history matters' is to analyze the rules that govern the respective explanatory games.

PHILIP: But there are many types of explanatory games, right?

STUDENT: Right. At every moment of time there are many explanatory games unfolding in parallel: mythical explanatory games, religious explanatory games, scientific explanatory games.

PHILIP: And I guess you want to say that the collective learning in each type of explanatory game makes it historical.

STUDENT: Yes, history develops according to basic rules that set the trajectory for development and that the participants learn in their socialization process.

© The Author(s) 2018
C. Mantzavinos, *A Dialogue on Explanation*, SpringerBriefs in Philosophy
https://doi.org/10.1007/978-3-030-05834-0_3

PHILIP: I guess in the case of scientific explanation, this occurs in processes whereby *apprentices* learn the way to play the game from *veterans* in the respective scientific community.

STUDENT: Indeed. And in the case of religious explanatory games, it occurs in the process whereby *deacons* learn the rules of the explanatory game from *priests* in the respective religious community.

PHILIP: And there is a cognitive division of labor.

STUDENT: Yes, and what concerns us here more specifically, a cognitive division of explanatory labor. Explanatory process is fundamentally social: some provide the big intuitions; others provide the appropriate elaboration of the means of representation; others build bridges to general theoretical structures; and still others work out the empirical testing and applications to different range of phenomena.

PHILIP: This is the contrasting view to that of the solitary explainer.

STUDENT: And to the view that the production and testing of explanations is a species of exclusively theoretical reasoning where strings of symbols are contrasted with other strings of symbols.

PHILIP: But you have got to be more specific!

STUDENT: Let me give you an abstract characterization of the rules of an explanatory game. There are four kinds of rules. First of all there is a basic set of *rules that constitute the game as a game*: the rules determining what counts as an explanandum, the rules determining what must be taken as given and the rules determining the metaphysical presuppositions.

PHILIP: I see, metaphysics should also be part of that analysis.

STUDENT: Yes, no explanatory game can take place in a metaphysical vacuum. The structure of the game is predicated on prior assumptions concerning the way the world is and the means by which it is explainable in principle. These rules can vary from stone-age metaphysics to highly refined metaphysical assumptions.

PHILIP: And the rules determining what is to be taken as given delineate the background knowledge, I guess?

STUDENT: Exactly. And the rules determining what counts as an explanandum belong also to the constitutive rules. Aristotle famously regarded motion as an explanandum and he provided causes as an explanans. Newton did not regard motion as an explanandum, but the change of motion—the games are different.

PHILIP: So, the *constitutive* rules seem to be straightforward. What other kinds of rules would be important?

STUDENT: The *rules of representation*.

PHILIP: Like graphs, scale models, computer monitor displays, etc. but also more abstract, I guess, like mathematical models?

STUDENT: Yes, these are the rules of representation characterizing scientific explanatory games. But remember science is not the only game in town! Because of the hospitality the people of Attica showed Demeter when she was searching for Persephone, the Olympian goddess first gave agriculture to the human race. This mythical explanation can be represented—in a quite primitive way—just by means of an oral story or in written form by the use of metaphorical language. But it can also be represented with the aid of some visual representations like a fifth-century

BCE red-figure attic stamnos, showing Triptolemos teaching agriculture, the gift of Demeter.

PHILIP: Besides the constitutive rules and the rules of representation, which other rules do you have in mind?

STUDENT: *Rules of inference*. These can be some rules of logic, but also a great array of other inferential strategies, which aim at whatever is regarded as the explanandum. They can be very diverse, ranging from recipes for formulating predictions to instructions about how to construct arguments.

PHILIP: And these rules of inference act on the representations, I guess.

STUDENT: Yes. I would say that in scientific explanatory games, laws or law-like statements along with some formal logical requirements are the basic rules of inference.

PHILIP: So, laws as inference-tickets, like the good old-fashioned instrumentalists!

STUDENT: Not necessarily so. We do not need to commit ourselves to a specific stance towards lawhood. All that I want to say is that some kinds of inferences are acting on representations whenever we provide explanations. In everyday explanations—but also in scientific ones—analogy is a powerful and extremely frequent inferential rule. In religious explanatory games, on the other hand, inferences are frequently made from experience of revelation to phenomena easily observable without sophisticated means of scientific representation.

PHILIP: And where are all these rules supposed to apply?

STUDENT: This is something that the *rules of scope* mandate. These are rules of specification, i.e. they give instructions about the scope of phenomena to which the explanatory game should be applied.

PHILIP: I guess a good example would be the explanatory game that evolutionary theorists were originally participating in—the initial scope included plants and animals. The scope has been extended then to include human culture.

STUDENT: And it has been extended and specified further to include even routines employed by firms in market competition and much more.[39] You see, one main property that I am interested in is *nestedness*. The rules of scope give instructions on how to dock one explanatory game into another and so provide nested games. The easier the fit between the rules of scope of different explanatory games, the higher the degree of their nestedness and the greater the potential of interlocking different explanatory games.

PHILIP: So, it is easier to get into a game, if nestability is given. You need not learn a lot of new rules in order to get into it.

STUDENT: Nestedness is my analogon to depth in the reductionist accounts of explanation.

PHILIP: Now, give me a concrete example!

STUDENT: Take the explanatory game that the classical economists were playing in the 18th century. The constitutive rules included those that determined (a) the value in exchange as the explanandum. In the *Wealth of Nations*, Adam Smith's aim was 'to investigate the principles which regulate the exchangeable value of commodities.'[40]

Then there were the rules determining (b) the bedrock of unquestioned facts and beliefs which were left out of the explanatory game in an implicit manner, and (c) the metaphysical presuppositions, i.e. that the social world is in principle knowable, that it exemplifies an order commensurate to the order of nature and that it is governed by laws which are discoverable.

PHILIP: The rules of representation were very basic, I gather: the explanation of the exchange value of commodities employed natural language and some numerical examples.

STUDENT: Yes. The rules of inference were the rules of logic prevailing at that time along with a series of law-like statements, the most important one being included as the title of Chap. 1, Sect. 1 of Ricardo's Principles: 'The value of a commodity, or the quantity of any other commodity for which it will exchange, depends on the relative quantity of labour which is necessary for its production, and not on the greater or less compensation which is paid for that labour.'[41]

PHILIP: And the scope of this objective theory of value founded on labour was both 'the early and rude state of society' as Adam Smith called it and 'the improved societies'. I particularly like the characteristic phrase: 'If among a nation of hunters, for example, it usually costs twice the labour to kill a beaver which it does to kill a deer, one beaver should naturally exchange for or be worth two deer.'[42]

STUDENT: More generally, I would say that the rules of scope of this explanatory game referred to conditions of competition operating without restraints. These rules enabled this explanatory game to become embedded into another explanatory game that had as an explanandum the value of the factors of production. The explanatory game of the exchange value of commodities is nested into the explanatory game of the value of factors of production.

PHILIP: So, the explanatory game was played by the classical economists following grosso modo those rules.

STUDENT: Yes. Within these rules their explanatory activities unfolded. Smith defended the doctrine that outlays on wages determine relative prices, and Ricardo attacked this doctrine, charging Smith that his measuring rod, the purchasing power of commodity over labour, would not do. A labour-embodied approach that Ricardo favored was opposed to a labour-commanded one.[43] He thought that the amount of labour that a product can command in exchange constitutes a poor measure of value. Malthus for his part was very critical of viewing labour, and corn as the measure of labour, as the measure of value.[44] And so on.

PHILIP: And there was a cognitive division of explanatory labour, I guess.

STUDENT: Adam Smith provided the big intuitions, David Ricardo specified more rigorously the rules of inference, and John Stuart Mill extended the scope of applications of the game.[45] A flow of explanations was produced and criticized, but mainly on theoretical terms: neither experimental evidence nor statistics were employed, nor any detailed cases from economic history were available then.

PHILIP: So, the change of the rules of explanatory game encapsulating the change from the 'objectivist' to the 'subjectivist' account of value a century later was not prompted by experimental evidence?

STUDENT: No. It was mainly theoretical criticisms that gradually led to the change in the rules of inference. And it was the introduction of mathematical models that gradually led to a new set of rules of representation. And the rules of scope also made it possible to extend the analysis to conditions of monopoly and later oligopolistic competition.

PHILIP: All kinds of rules have changed then: the rules of representation, the rules of inference and the rules of scope.

STUDENT: Indeed. The change was initiated by the marginalists, Walras,[46] Jevons[47] and Menger[48]—but they were all still playing the same explanatory game since the constitutive rules had not changed: the explanandum remained the value of commodities and factors of production, the metaphysical assumptions remained largely the same, and a great part of what was taken for granted by the classical economists was also taken for granted by the marginalists.

PHILIP: I think the most important change must have been in the rules of inference, since the maximization principle was introduced. This provided greater generality and unification since it made it possible to explain both factor and product prices with the help of a single principle.

STUDENT: That was indeed the case. Classical economists derived the prices of products from the 'natural' rates of reward of the three factors of production, which were in turn explained by three separate principles: wages of labour were determined by the long-run costs of producing the means of subsistence; land rentals were determined as a differential surplus over the marginal costs of cultivation; and the rate of profit on capital was regarded as a residual.

PHILIP: And the maximization principle also facilitated the introduction of new rules of representation, i.e. mathematical calculus, more specifically maximization under constraints.

STUDENT: Exactly. The maximand can be either utility or physical product or profit, but the formal representational rule remains the same. The application of this formal means of representation has made the explanations more precise, simple and offered a unification of diverse economic phenomena. Finally, the scope of the explanatory game has been extended to include also conditions of monopoly rather than only competitive conditions.

PHILIP: So, all kinds of rules have changed, the rules of inference, the rules of representation and the rules of scope—all without any kind of experimental evidence invoking such a change!

STUDENT: The advantages of the adoption of the new rules were mainly conceived by the economists of the era in terms of their increased consistency, increased accuracy and simplicity. And the unification provided by the extension of the scope of the explanatory game was also regarded as an additional advantage.

PHILIP: This is a very interesting case. But, you know, not many people regard economics as an exemplary science!

STUDENT: Then let me give you an example of an explanatory game in medicine—this has always been regarded as a science par excellence!

PHILIP: This is a very strange claim! Medical science has always been full of quacks. And in medicine, it's routine that what's taken as gospel today, could be reversed tomorrow.

STUDENT: Stop being ironic!

PHILIP: Just to give two quick examples: whether menopausal women should take estrogen or whether wine is good for you!

STUDENT: I will give you an example which is as uncontroversial as it is important.

PHILIP: I am listening.

STUDENT: Take the explanatory game that concerns the functioning of the heart and the circulation of the blood. This is an interesting case for two reasons: first because of the quite spectacular fact that a change of the rules of the explanatory game took about thirteen centuries to unfold and second because of the role of the institutional framework in which the explanatory games themselves are embedded.

PHILIP: I guess that this explanatory game starts with Galen at the end of the 2nd century ACE and you say that the rules changed fifteen centuries later—with Harvey?

STUDENT: Yes, and this is a history of an incremental change rather than a revolution. Starting with the constitutive rules, (a) what counted as an explanandum was an unambiguously delineated phenomenon: blood circulation and the functioning of the heart, (b) the context of background knowledge remained implicitly unquestioned, and (c) a series of metaphysical presuppositions were very important: a teleological worldview according to which 'Nature does nothing in vain' as Galen characteristically stresses in his *De Usu Pulsum*,[49] and, of course, the postulation of *pneuma*.

PHILIP: Yes, the idea of a 'life-giving spirit' was a constitutive part in nearly all medical explanations of the era.

STUDENT: The unique character of the rules of representation is interesting: the description of the phenomenon has largely been obtained by means of dissection, which allowed a direct representation of the heart and the blood circulation by all five senses: sight, hearing, touch, smell and taste!

PHILIP: Though I guess that the use of natural language remained also among the options!

STUDENT: The rules of inference used were mainly the rules of logic and those of analogical reasoning: since dissections of human bodies were not among the options (though they did see inside humans in surgery and by chance), the rule employed was that the cardiovascular system of animals functioned in an analogous way to the cardiovascular system of humans.

PHILIP: And the scope of the explanations was straightforward, this very cardiovascular system.

STUDENT: Indeed. Now, the institutional rules within which the social interactions unfolded in that historic period are also important. A feature of medicine as it was practiced at that time was that several doctors were often summoned to the patient's bedside, where they offered competing diagnoses and prognoses, and it was up to the patient himself or his representatives to choose among them.[50] The academic reputation was partly earned in public disputes where rhetorical skills, among other, were essential for success.

PHILIP: These were times where the institutions guiding scholarly research and teaching were entirely different than in the Middle Ages and of course in the modern era!

STUDENT: Addressing the question of how blood gets from the arteries to the veins he suggested that blood permeates from pulmonary arteries to pulmonary veins through invisible channels. The resulting blood in the pulmonary veins does not reach the left ventricle, but it is rather used as nourishment by the lungs; hence there is no pulmonary circuit. Instead, his main inference was that blood in the left ventricle, and so the systemic arteries, is derived directly from the right ventricle, through invisible pores in the interventricular septum.[51]

Although Galen did recognize the unidirectional characteristic of the cardiac valves, Galen was not able to explain the circulation through the heart since he adhered to the view that blood was subject to an ebb and flow motion. This remained the main, more general, *inferential rule*.

PHILIP: The notion of the pump was not available at that time, consequently *drawing such an analogy* and inferring that also the heart functions as a pump was *not among the possible options*.

STUDENT: Yes, Galen inferred instead that the forward flow of blood from the right ventricle through the lungs was due to the rhythmic respiratory movements of the thorax. The explanations of the cardiovascular system remained inaccurate for various reasons: specifically, due to the *metaphysical rules* employed that involved pneuma as an ingredient in the blood circulation; due to the *rather limited rules of representation based solely on anatomical means*—dissection of dead animals and vivisection of living ones; and due to the *inferential rules* employed which, with the exception of the logical ones that secured the validity of argumentation, did not provide accuracy.

PHILIP: But there must have been controversies with respect to all kinds of rules. The 'Dogmatists', the 'Empiricists' and the 'Methodists' held very different views about medicine![52]

STUDENT: Of course. Empiricists denied the epistemological merit of dissection, for example. But, despite the controversies, by 600 ACE at the latest, Galen's explanatory practices had turned into Galenism: one individual's opinions had become an intellectual system that guided all medical learning.[53] Challenging the views of Galen remained unthinkable for centuries. Indeed, his authority was so formidable that for more than a millennium the dictum was: if there proved to be no holes in the septum, it clearly followed that nature must have undergone changes since Galen![54]

PHILIP: How could a change in the rules come about then?

STUDENT: Though there were isolated critical voices, it was a novel kind of institutional framework that gradually emerged that ultimately enabled the *institutionalization of criticism* by medical scholars and that in turn led to an incremental change of explanatory rules and an abandonment of those of Galenism.

PHILIP: But what do you mean by an institutional framework? This is very abstract and vague!

STUDENT: An institutional framework comprises the *informal* and *formal* social rules that structure human interaction. *Informal institutions* are made up of *conventions*, *moral rules* and *social norms* and are enforced by the other members of the group and in the case of the moral rules by what we call 'conscience'. The *formal institutions* are enforced by the state—they are the *legal rules of the society*. In this broad picture the intricate mix of informal and formal institutions shape the different patterns of social interaction. Besides individuals, it is also organizations (bound by specific rules designed to accomplish their aims) that are the actors and that ultimately give rise to the diverse social patterns.

PHILIP: And how does all this connect with what we are talking about now?

STUDENT: The informal institutions encapsulate the critical attitude which, having started in Ancient Greece, was weakened through the centuries, was somehow revitalized with the Condemnation of 1277[55] and was revived again during the Scientific Revolution. The incremental spread of an anti-dogmatic attitude during the Renaissance and the invention of printing press in 1450s provided the possibility of the rapid dissemination of medical knowledge and helped human creativity to flourish—and more narrowly it also aided in the domain of the explanations of the cardiovascular system that concern us here.

PHILIP: And what about politics?

STUDENT: The *formal part of the institutional framework* which had to do with the prevailing political institutions was also conducive to allowing criticism and to the unimpaired unfolding of the inquisitive spirit of medical scholars. This was the case in the Republic of Venice, of which Padua was the university from 1405, and which largely guaranteed civil and religious freedom and tolerance. Venice was the most anti-clerical state in Europe. Finally, the *organizational structure* of the University of Padua seemed to have been pivotal not only for scientific discoveries in medicine, but also for other disciplines during the Scientific Revolution, since also Copernicus and Galileo were at this university during their most productive periods.[56]

PHILIP: Avoidance of nepotism and clientelism was characteristic of the organizational structure of the University of Padua—this was indeed the case. And one specific organizational rule that has always impressed me was that, although every faculty member was allowed full freedom to teach, he was not allowed to repeat his course of lectures in successive years.[57]

STUDENT: Andrea Vesalius, who was a professor at Padua, provided a critical innovation concerning the *rules of representation* with the publication of his *Fabrica* in 1543.[58] He collaborated with the workshop of Titian to construct wood block engravings of the drawings contained in the *Fabrica*. These anatomical drawings and illustrations were the outcome of a productive synthesis of art and science and reflected a naturalism in the depiction of the human anatomy that was radically different than the conventional medieval drawings.

PHILIP: And the printing technology made a dissemination of diagrams and drawings of increased accuracy possible...

STUDENT: The representation of the explanandum phenomena became, thus, more precise and more easily accessible. The direct representation by means of the senses in the process of a dissection was replaced by the printed representation of the heart and the other organs by means of drawings and other illustrations. The successor of Vessalius in Padua, Realdo Colombo, was able to provide an explanation of what is called the smaller circulation, i.e. the passage of the blood from the right side of the heart into the lungs and from there into the left ventricle of the heart.[59] And later, in 1571, Cesalpino for the first time mentions the 'circulation of the blood', which was supposed to be confirmed by evidence derived from dissection.[60]

PHILIP: But wasn't it Harvey who provided the explanation of blood circulation?

STUDENT: This is exactly the productive function of employing the notion of an explanatory game! The debate among generations of medical historians about whether Cesalpino or Harvey should be credited with the 'discovery of blood circulation' comes under a new light. Explanations are constantly produced and criticized, and they are the outcomes of a historical process during which the rules within which the games unfold are constantly changing. The *rules of inference* that Cesalpino used to draw his conclusions and provide his explanations were the same ones used before him: these were the general inferential rules leading from observations reached by the use of his senses to more general conclusions. This was not the case with Harvey, who was the first to use an arithmetic example, and thus a quasi-mathematical rule of inference for the first time in the provision of his explanation of blood circulation: according to a quite quick calculation by Harvey, the sheer quantity of blood sent from the left ventricle to the aorta was so extraordinarily big that it became reasonable to hypothesize that blood circulates. But the novel rules of inference that Harvey employed also included the analogy of the heart as a pump, which decisively differed from the Galenic view of blood, which was thought to ebb and flow in the arteries.

PHILIP: I see—your argument is that it was the integrated employment of all those rules, some of them adopted from the past use, some created by Harvey himself, that led to a novel explanation of the blood circulation.

STUDENT: Exactly. All rules had changed in the meanwhile, including the metaphysical ones, since the postulation of different spirits common in the traditional Galenic framework had been abandoned. Besides, the *communication* of the explanation of Harvey was provided in a more systematic and synthetic fashion by means of his short monograph—*Exercitatio anatomica de motu cordis et sanguinis in animalibus*.[61] It focused on the issue rather than dealing with the problem within the framework of general treatises. This was more *persuasive* to his peers and more easily acceptable.

PHILIP: At that time the microscope had already been invented and used, right?

STUDENT: Yes, and this allowed a crucial refinement of the rules of representation. It was Malpighi's observation of the lungs of frogs with the aid of a microscope in 1661 that enabled capillaries to be identified; consequently, the passage of the blood from the outermost branchings of the arteries to the outlying ramifications of the veins could be shown.[62]

PHILIP: All of that makes sense, but in the end how active was the role of institutions in the unfolding of this explanatory game?

STUDENT: Institutions are there to canalize human passions. And human passions also rule the souls of scientists. The acceptance of the use of novel rules of representation by way of microscopic studies was anything other than straightforward: although Malpighi was a very modest and gentle man, the immense scope and impact of his microscopic studies provoked such envy and criticism that in 1684 his villa was burnt down by adversaries, his papers, notes and manuscripts destroyed and his laboratory equipment ruined!

PHILIP: Human passions, the guide of all human activities!

STUDENT: And naturally also of explanatory activities. In the traditional account, which I reject, explanations are abstracted from their dynamic context of inquiry, of observation and complex reasoning and treated as outcomes, as elements in a static, unchanging construction, as museum pieces. In the account that I propose, explanations are an organic part, a provisional outcome of ongoing explanatory activities undertaken in diverse social contexts. Institutions are the social normative rules that channel these social contexts and give them a semi-crystallized shape. Miguel Serveto, in the fifth book of his religious treatise *Christianismi, Restitutio*[63] described the small circulation quite accurately in an account of the passage of the blood from the heart to the lungs and then back again to the left ventricle of the heart. However, this work, though published already in 1553 and obviously based on studies that took place earlier—indeed, earlier than Colombo's *Re Anatomica*—had no impact whatsoever in the explanatory process of the functioning of the heart and the blood circulation: having been found offensive to Catholics, Lutherans and Calvinists, the author was condemned for spreading and preaching Nontrinitarianism and sentenced to death by burning at the stake in Geneva in 1553 with one of the apparently last copies of his book chained to his leg. Of 1000 copies of the book, only three still exist today—the one in Bibliothèque National Paris, most likely used in his trial. What does this example show us then? It shows us that metaphysical assumptions that were irrelevant to the explanatory game were backed up by institutional power, and the result was that the explanatory process was inhibited by means of the physical extinction of both the one engaged in the explanatory activity and the product alike.

PHILIP: I certainly agree that institutions are eminently important as you say.

STUDENT: And it is important to stress also that at every moment of time there are many explanatory games taking place concurrently. This is particularly interesting for both descriptive and normative purposes, since we can identify different explanatory games that deal with the same explananda, but which differ from one another on the basis of what they assume as given and of their metaphysical assumptions. Three important types here are *mythical*, *religious*, and *scientific* explanatory games.

PHILIP: Cosmogony, the emergence of the universe, would be a nice example, I think.

STUDENT: The Greek mythical explanatory game produced explanations of the emergence of the universe like that of Hesiod's Theogony[64]

Tell me these things, Olympian Muses, tell

From the beginning, which first came to be?

Chaos was first of all, but next appeared

Broad-bosomed Earth, sure-standing for all…

PHILIP: And the Christian explanatory game produced probably the most popular explanation of cosmogony as it is contained in the Book of Genesis[65]:

'In the beginning when God created the heavens and the earth, the earth was a formless void and darkness covered the face of the deep, while a wind from God swept over the face of the waters. Then God said, "Let there be light"; and there was light. And God saw that the light was good; and God separated the light from the darkness. God called the light Day, and the darkness he called Night. And there was evening and there was morning, the first day.' And so on for the rest six days...

STUDENT: Compare these with the contemporary cosmological account of the emergence of the universe based on a big bang explanation. According to the general theory of relativity, gravity is not a force, but the curvature of space time. The range of gravity is infinite since it is a property of space time itself and the evolution of the universe is ultimately dictated by gravity. A set of differential equations relates the dynamic quantities in the universe and turns them back in time. Simulations on computers relate initial conditions of the universe to different outcomes aiming at representations of the kind of universe we live in. Earlier astronomers, most prominently Edwin Hubble, observed that neighbouring galaxies of the Milky Way galaxy are receding away—the more distant they are, the faster were they found to be moving away. In other words, recessional velocity of a galaxy increases with its distance from the earth. Reversing this expansion scenario back in time, the inference is drawn that if galaxies are moving rapidly apart now, they must have been denser—with more matter and energy per unit volume—in the past. Going back in time, the whole universe should converge to a point of infinite density and extremely high temperatures, and this should be the starting point of big bang. The universe itself should have been an infinitely dense point which expanded to its present size. The age of the universe according to this Big Bang explanation is estimated to be 13.73 billion years.

PHILIP: With all these examples the notion and function of an explanatory game have become much clearer. But I wonder whether this approach can entirely avoid posing and answering what you have called 'the wrong question', i.e. 'what is an explanation?'

STUDENT: Have you heard me doing so?

PHILIP: No. But the issue is whether your purposefully avoiding it can be maintained to the end.

STUDENT: I think that we can speak and argue intelligibly on a great range of issues and problems based solely on implicitly shared understandings.

PHILIP: Sometimes, but not always, and probably not in this case.

STUDENT: Let me show you something. We have to walk a few steps up to the Temple of Poseidon. Here it is!

PHILIP: B y r o n !

STUDENT: He inscribed his name here, possibly during his first visit to Greece. An attempt to remain in eternity, probably!

PHILIP: This is a charitable interpretation of this inscription!

STUDENT: But there is no evidence that it was he who carved his name into the base of one of the columns of the Temple.

PHILIP: But this is probably the best explanation.

STUDENT: Perhaps the name of Lord Byron was carved by an enthusiast admirer rather than by the poet himself. In any case, he did dedicate some verses to Sounion:

Place me on Sunium's marbled steep,
Where nothing, save the waves and I,
May hear our mutual murmours sweep;
There, swan-like, let me sing and die;

PHILIP: Wonderful verses! But tell me, what kind of activity is required here, *explanation* or *interpretation*?

STUDENT: It is clearly an interpretation of an inscription that we are engaging in!

PHILIP: But this presupposes that we have a clear view of the difference between explanation and interpretation.

STUDENT: It sure does. But to avail of a clear view does not presuppose in turn that we avail of precise definitions of the two concepts.

PHILIP: I am not entirely convinced, but let us walk down the hill now.

STUDENT: Yes, let us walk down the hill, and I will show you something that will surprise you.

Part IV
On Explanatory Progress

STUDENT: In the meanwhile, let me say a few more words about the normative side of this philosophical account of explanation—unless you think that we have exhausted the subject?

PHILIP: No, the normative dimension is probably the most important one. I particularly used to like Peter Railton's notion of an 'ideal explanatory text'.[66]

STUDENT: But this is an attempt to establish an eternal standard for judging the quality of explanations. It is so far away from the real-world explanatory practices that is in the end useless!

PHILIP: First of all, ideals need not be useful to be valuable. Further, you can never proceed to the formulation of normative judgments without some kind of an ideal.

STUDENT: You are right, of course, with respect to your first point. Regarding your second point, I would only partly agree: normative guidance is required for any kind of activity and also for explanatory activity. Such guidance can, but need not invoke eternal ideals and even less so single ideals. Multiple values can guide explanatory activities, and I claim that they in fact do so.

PHILIP: So, value pluralism is the key notion here.

STUDENT: Yes. And a fine-grained normative account is enabled, thus. We can easily evaluate the different types of rules of any explanatory game with respect to different values.

PHILIP: But would not that be too complex?

STUDENT: Not necessarily; all that is needed to organize our thoughts is to think in a multi-dimensional framework. Start with a simple hierarchical model, with *problems* at the *lowest* level, *rules* at the *intermediate* level and *values* at the *highest* level.

PHILIP: I guess you suggest that we should favor *problems* rather than *facts* as the starting point.

STUDENT: Exactly. Explanatory activity can easily be conceptualized as a problem solving activity during which the goal state is a target to be explained. Problems

© The Author(s) 2018

C. Mantzavinos, *A Dialogue on Explanation*, SpringerBriefs in Philosophy
https://doi.org/10.1007/978-3-030-05834-0_4

are very different from facts, and solving a problem cannot be equated with accounting for a fact.

PHILIP: I like this a lot: pragmatism commonly views problems as the starting point, but so does Popper who famously claimed that *All Life is Problem Solving.*[67] And Laudan has ingeniously made the point that it would be extremely difficult, if not impossible, to account for scientific activities by starting with facts, for the simple reason that scientists very often are engaged with issues that cannot be regarded as states of affairs.

STUDENT: Whether angels are male or female, for example, or the problem of describing the exact properties of a socialist utopia!

PHILIP: And Laudan has this nice phrase: 'a problem need not accurately describe a real state of affairs to be a problem: all that is required is that it be thought to be an actual state of affairs by some agent.'[68]

STUDENT: And there is a rather simple point, though seldom acknowledged: many facts about the world do not present themselves as problems simply because they are unknown. And the very question of what constitutes a problem is very often a contestable issue, which is solved by appeal to the rules of the explanatory game—as are most divergences of opinion about how a specific explanatory problem is to be solved.

PHILIP: So, moving one level up in the hierarchy is the first move.

STUDENT: We appeal to rules of representation, rules of inference and rules of scope whenever explanatory problems are at issue. But none of these rules is sacrosanct, of course. So, there is analysis, debate and criticism of the rules themselves.

PHILIP: And I guess this debate and criticism takes place with respect to diverse values—that is by moving one further level up in the hierarchy?

STUDENT: Exactly. The criteria of appraisal encapsulate diverse values, i.e. epistemic values like accuracy, simplicity, consistency etc.,[69] and non-epistemic values, i.e. aesthetic, moral, political, religious values like beauty, honesty, freedom of expression, piety, etc.[70]

PHILIP: The axiological level of values is the highest level in *this* hierarchy, I see.

STUDENT: Now let me complicate the picture a bit: the multiple rules and the multiple values enable a multidimensional critical appraisal. Values provide fixations of normative resources that can help us judge the goodness of explanatory rules that have emerged in historical time while the explanatory games have developed. And most importantly, values provide the normative resources for judgments across explanatory games that are permanently unfolding in parallel.

PHILIP: What does the multidimensionality consist in?

STUDENT: It consists in the possibility of evaluating different types of explanatory rules with respect to different values. Rules of representation, for example, can be evaluated with respect to accuracy or simplicity; rules of inference with respect to consistency, accuracy or simplicity, etc. A certain rule can be evaluated positively with respect to one value, but negatively with respect to another.

PHILIP: I can see how such an evaluation would work with respect to epistemic values. But how do the non-epistemic values enter the picture?

STUDENT: Explanatory games are always embedded in a framework of formal and informal institutions. Remember the case of Miguel Serveto! The evaluation of the institutional framework of his time with respect to the non-epistemic value of freedom of expression, for example, is clearly a negative one. And similarly with the institutions of the social scientific explanatory games in the Soviet Union, to take another extreme case: Whatever explanation the Communist Party did not like, it could not survive.

PHILIP: But are not some values more important than others? Many philosophers regard truth as the most important value!

STUDENT: But in many religious discourses truth is also regarded as the most important value. In Christianity, for example, Jesus *is* the Truth. And since I want to encompass all explanatory activities in my framework, i.e., also religious ones, among others, unfortunately, the notion of truth cannot function as a bridge. From the point of view of the participants in a religious explanatory game who commonly use the notion of truth to describe the main property of a deity, debates about versions of the correspondence theory of truth must seem as artificial puzzles regarding aesthetically pleasing irrelevancies.

PHILIP: You are probably right about that. But still, to give my concern another twist, the question remains: Is there some kind of external criterion or justification process which would help us claim that one value that is used in the critical appraisal of the rules of the explanatory game is superior?

STUDENT: No, there is not. It would be very nice, if we had such a criterion or if we could devise a process that would guarantee a certain result. But we are still in the situation that Otto Neurath described in an unsurpassable way: 'Wie Schiffer sind wir, die ihr Schiff auf offener See umbauen müssen, ohne es jemals in einem Dock zerlegen und aus besten Bestandteilen neu errichten zu können.'[71]

PHILIP: Das ist so, weil wir sehr unvollkommene Wesen sind.

STUDENT: The quest for an ultimate justification is a manifestation of the vain quest for certainty originating in the idea of a positive, sufficient justification. The German philosopher Hans Albert called it the *Münchhausen Trilemma*. The demand for a justification for everything leads to a situation with three alternatives, all of which are unacceptable. One has to choose between an infinite regress, a logical circle in the deduction, or a breaking-off of the process at a particular point, which can always been done in principle, but involves an arbitrary suspension of the principle of sufficient justification.[72]

PHILIP: But I thought that we are familiar with this unfortunate situation already in the context of ancient skepticism—isn't it a version of the discussion of Agrippa's five tropes?[73]

STUDENT: Indeed, but Hans Albert has shown how this situation can be avoided: by substituting the principle of critical examination for the principle of sufficient justification. In accord with this principle, a discussion in ethics, philosophy of science or epistemology need not refer back to ultimate reasons in order to be 'rational'. Instead, problems that arise in the sphere of cognition and in the sphere of praxis are to be discussed and solved in light of the already available solutions. Solutions to entirely new problems require creativity and imagination, and the important requirement is the availability of criticism, in order that we can learn individually and collectively from our errors. This is why an appropriate institutional framework is vitally important!

PHILIP: The role of institutions again!

STUDENT: Indeed. It enables discussion and criticism of alternative solutions and, thus, ultimately the possibility to learn from errors. This encapsulates better the requirement of rationality, I would say. Solutions are not judged as good or rational, because they resemble or come closer to a predefined ideal nor because ultimate reasons can be given in their favor. All our solutions are fallible, but we can improve them by critical discussion.

PHILIP: But how can one apply this approach to problems of explanation?

STUDENT: Explanatory methodology can be conceptualized as a rational heuristic. Instead of debating whether a certain explanation fits or does not fit a specific ideal, a comparative evaluation of a multidimensional character can be performed. Explanatory methodology can thus be viewed as a technological discipline.

PHILIP: How exactly?

STUDENT: A technology operates with hypothetical rather than categorical imperatives. To judge the quality of different sets of rules of an explanatory game, one needs only to hypothetically presuppose certain values and criteria and then to critically inquire whether the explanatory activities guided by these rules can fulfil these criteria—say accuracy or empirical fit. If other values are regarded as important, one can have an equally rational discussion about whether the same rules are valued positively with respect to, say, beauty.

PHILIP: So values are not excluded from the discussion, but make up an organic part of it.

STUDENT: Yes. We can disagree about which value is to be considered important, but still we can have a positive discussion about which rules attain these values without endorsing them. The technological character of explanatory methodology consists in acknowledging that the debate can be of the standard genre of a normal means-ends framework: rule X (of type a), compared with rules Y and Z (of the same type), is more accurate; rule A (of type b), compared with rules B and C (of the same type), is simpler, etc.

PHILIP: Give me an example of how this is supposed to work in practice!

STUDENT: Take the three different explanatory games dealing with cosmogony, the *mythical*, the *religious* and the *scientific*. One can proceed to unambiguous judgments about the degree of accuracy of the rules of representation used in the three explanatory games. The narratives in Hesiod's Theogony and in the Book of Genesis are clearly less accurate than the algebraic formulas and computer graphs used in modern cosmology. This judgment is independent on whether one thinks that accuracy is an important value or not and also on what stance one takes regarding the metaphysical presuppositions of the three games.

PHILIP: So, the hypothetical imperatives employed are of the type: if you want to attain an accurate explanation of phenomenon A, then you should employ rules of representation B and C rather than X or Y.

STUDENT: Exactly.

PHILIP: Can one give arguments for them?

STUDENT: Of course. Arguments in favor of this and other hypothetical imperatives are to be proposed and debated in a process of critical discussion!

PHILIP: I see.

STUDENT: This conception of explanatory methodology can also be conceptualized as a rational heuristic. For many decades, indeed centuries, the discussion has been centered on questions of validity. This was the case in general epistemology in the original discussions around the difference between δόξα and ἐπιστήμη, belief and knowledge. The implicit presupposition of this discussion was that a criterion or process could be ultimately found that would identify a class of beliefs as true knowledge; knowledge as justified true belief is still the account that a series of epistemologists favor. And in philosophy of science the discussion took a new turn when one stopped asking about the origins of scientific knowledge, and started asking about the validity of scientific knowledge. The famous distinction between the context of discovery and the context of justification had crystallized the two families of problems. And for a very long time the context of justification was regarded as the only legitimate domain of the philosophy of science. Issues of scientific discovery were pushed out towards psychology or even worse irrationalism.[74] But this was an unfortunate development, I think.

PHILIP: I think that this very distinction is not helpful.

STUDENT: I would disagree. I think it is helpful, but we should say something sensible about the *ars inveniendi* and not only about the *ars judicandi*.[75] What I have in mind, more specifically, is this: Explanatory methodology should be practiced as a rational heuristic, as a discipline paying attention to both the genesis and the evaluation of explanations along the lines that I have suggested.

PHILIP: Where does the heuristic element lie in what you suggest?

STUDENT: In the exercise of imagination when proceeding in the evaluation of the different explanatory rules with respect to the diverse values and in the exercise of a creative choice which will favor one or more of the proposed alternatives.

PHILIP: Imagination and creativity should be an integral part of every methodology, is this what you suggest?

STUDENT: To state the same idea differently, let me just provide the contrast—and this is the automatic application of a worked-out normative model on a specific subject matter. The main point is this: we have made a first progressive step by proposing and accepting the view that there is no algorithmic form for generating new certain knowledge. However, this important step has two repercussions. The first is that one excluded heuristic from scientific methodology altogether—the suggestion was that since there is no algorithm enabling the attainment of new scientific knowledge, then there is no point of designing a heuristic for science. The second is that the quest for certainty was transferred to the evaluation of the products of scientific endeavor, i.e., hypotheses, explanations, etc.—if one can find an appropriate algorithm for evaluation, then one still has the key to accepting genuinely valid scientific knowledge.

PHILIP: And there has been a sustained effort to construct formulas for evaluating explanations.[76]

STUDENT: These are futile endeavors to replace controversies with calculations. There is no algorithm relieving us from the necessity of making choices. Controversies cannot be substituted with calculations—we still have to form judgments and decide which calculations to adopt.

PHILIP: I agree. So this speaks to the heuristic element of the approach that you suggest, but what about the rational element of a 'rational heuristic'?

STUDENT: An explanatory methodology is not to be viewed as rational because it sets out to prove or to defend a specific algorithm or justificatory procedure as the most adequate one to solve the problems at hand. It is rational because it lays the ground for the construction of alternatives by specifying and clarifying the working properties of the different kinds of explanatory rules; and because it provides the basis for their criticism by highlighting the specific characteristics of a multidimensional evaluation of the different rules as they are embedded in the broader institutional framework. This is a *procedural rationality* which does not prescribe the content of the decisions that are ultimately made by the participants of the explanatory games themselves, but highlights instead the individual and collective conditions that must prevail in order for these decisions and judgments to be informed and to be adequate with respect to goals and values that they are aiming at.

PHILIP: So, this is how you want to account for procedural rationality?

STUDENT: Indeed. Dethroning Olympian rationality need not lead us to irrationalism or relativism. A rational heuristic of the kind I am suggesting can offer a guide to cognitive praxis without reducing explanatory methodology to calculations that do not leave any space for imagination and genuine decision making.

PHILIP: But there is a whole discipline, decision-theory, which offers a host of decision models—should we thoroughly discard it?

STUDENT: Most of these decision models have been created in the optimistic spirit of the second half of the twentieth century, during which *reason became rationality*. During this period the faculty of reason was radically reconceived by philosophers, economists, mathematicians and social scientists. The main move has been to replace self-critical judgments of reason with the algorithmic rules of rationality in the models of decision theory, game theory, artificial intelligence, etc.

PHILIP: But the background of this transformation lies, as my friend Raine Daston has shown, in the history of rules starting in the early Middle Ages and lasting until the mid-nineteenth century. The chief sense of regula/rule/Regel/règle derived from Regula Sancti Benedicti (6th century), laid down for monks and used as a moral precept about how to conduct one's life. In the early decades of the nineteenth century the meaning of rules as algorithms begins its rise to dominance in contexts such as the first effective calculating machines and the attempts to guarantee the logical solidity of mathematical proofs. By the early 1950s, the dream of reducing intelligence, decision-making, strategic planning and reason itself to algorithmic rules had spread like wildfire to psychology, economics, political theory, sociology and philosophy of science. It is a long, complex history of how reason became rule-bound and how the faculty of judgment as an essential component of reason has occurred.

STUDENT: I understand. This is a much longer and more complex historical process. My position is not to throw away all decision making models, of course, but certainly those that simulate an inexistent precision in conditions of pervasive uncertainty.

PHILIP: But there are decision models under risk!

STUDENT: But calculable risk and uncertainty must be rigorously differentiated. Frank Knight brilliantly elaborated on that already in 1921 in his *Risk, Uncertainty and Profit*.[77] The point is as simple as it is important: the prevailing conditions whenever a choice of rules takes place are characterized by *genuine uncertainty* rather than *parametric uncertainty*, where probabilities can be calculated and assigned to options.

PHILIP: It remains to be seen whether this is a useful distinction.

STUDENT: Choice cannot remain choice and its outcomes be predetermined.

PHILIP: Predicting choice is very hard, this point I grant. But is it impossible?

STUDENT: Choice requires the imaginative faculties of the mind and involves a creative element that must be acknowledged and must find its place in any methodology.[78] The outcomes of those decisions are themselves fallible—probably most of the time creative decisions with the aid of imagination are erroneous.

PHILIP: That they are fallible does not mean that they are unpredictable.

STUDENT: You seem not to be convinced. Not yet, at least! Let me elaborate more on this.

PHILIP: I am listening.

STUDENT: There is a second aspect of procedural rationality, which has to do with the collective conditions that must prevail in order that an established correction mechanism is in place. Institutionalizing the possibility of criticism is the best means to facilitate the correction of errors when scientific choices are involved. Anchoring the freedom of criticism in the institutional framework of the society is the *collective* condition that must prevail and that enables procedural rationality to manifest itself.

PHILIP: This remains quite vague!

STUDENT: Let me become even more specific. Scientific explanatory activities unfold within the intricate institutional framework of modern science. It is a historical contingency that the informal and formal institutions of modern science have come to prevail in a long evolutionary process in the West. On the one hand, the *informal institutions* encapsulating the critical tradition coming from Ancient Greek philosophy,[79] weakened in the course of many centuries and revived during the Scientific Revolution; and on the other hand, the emergence of *competitive political structures*[80] that have considerably increased individual freedom while allowing criticism without pernicious consequences for the critic. In the modern era an intricate institutional matrix has come to prevail in most parts of the world, which has further cemented the freedom of expression that naturally gives rise to a plurality of opinions and fosters the competition among different views. Moreover, the gradual historical development of the set of organizational structures characteristic of *modern universities* enabled the pooling of a great array of resources—intellectual and material—dedicated to the constant generation and criticism of solutions to abstract theoretical and practical problems.

PHILIP: This is interesting. Go on!

STUDENT: Scientific activities are, thus, embedded in these broader norma-tive structures of the modern world. What appears to distinguish them from the other activities unfolding in ordinary life (including the religious ones) is the rela-tive sophistication and systematicity by which empirical evidence is generated and assessed, something that is enabled by the social and cultural arrangements as they are encapsulated in the institutional framework of science. It is *the possibility of criticism* provided by this framework which acts as a corrective to the error prone problem-solving activities in which scientists, like ordinary people, are engaged.

PHILIP: What kinds of errors do you have in mind?

STUDENT: Errors ranging from fallacious mental models that do not give an accurate representation of the environment to fallacious inferences (including con-firmation biases, erroneous probabilistic calculations and much more) and errors regarding the range of phenomena toward which the scientific explanations are to be applied.

PHILIP: So, it is the institutions of science that permit scientists to circumvent their inherent cognitive limitations. This is your point.

STUDENT: Yes. Scientists can get around some of them by exploiting a vast array of tools offered by the modern institutional setting of science: journals, professional associations, peer-reviewing, etc. The possibility of criticizing each other's work is certainly provided. These features of the modern scientific enterprise ensure to a great degree, I think, that the collective outcome of the individual endeavors of scientists is characterized by a higher quality and reliability than individual mistake-prone efforts. The criticism that comes from the social environment of science and is offered to the individual scientist can affect the metaphysical presuppositions, the quality of the mental representations, the inferences and the chosen scope of application of the explanations that he or she offers.

PHILIP: So, there is criticism all the way down!

STUDENT: Notwithstanding the cognitive and institutional mechanics that might lead to a cementation of explanatory rules, what I was suggesting before was that there is always a *competition* between alternative explanations and models offered and techniques employed by scientists. They certainly stand in competition among them, and they stand in a *broader competition* with those offered in religion and in everyday life. There is always a plurality of alternatives to choose from, and this is important to keep in mind.

PHILIP: The *Myth of the Framework*,[81] as Popper called it!

STUDENT: It is indeed a myth, a myth that exaggerates a difficulty into an impossibility. The decision to accept one set of alternatives is always a decision to reject another set, and the judgment leading to that decision involves the comparison of the alternatives with nature and with each other. The *criticism* that is possible in the scientific enterprise is much more extended because of the prevalence of the *liberal normative dimension* in which explainers in science usually find themselves operating, which is due to the institutional framework in place.

PHILIP: I see. What you offer in the end is a version of relativism, I would say. A very sophisticated one.

STUDENT: But I am not a relativist!

PHILIP: You are not. But you sound like one!

STUDENT: I am a rationalist. My conception of procedural rationality is tied to pluralism[82] and fallibilism as I was arguing before. My position is a middle-ground position between dogmatism and relativism.

PHILIP: Let me pose a last query now. You seem to believe in the possibility of scientific progress, which I also share as you know, but from what you have said it is not obvious how explanatory progress is possible in a pluralistic framework.

STUDENT: In discussions about progress much confusion can be avoided, I think, if one acknowledges the simple issue that judgments about progress are parasitic on the specification of goals. Progress is a comprehensible notion only with respect to the attainment of a specific goal or aim. To the question 'Is there any progress?' the response can only be....

PHILIP: 'Progress with respect to what?'

STUDENT: Exactly. In our case, one can sensibly speak of progress with respect to the attainment of a specific value. Does a sequence of extensions, refinements and generalizations, say of the rules of representation R_1, R_2, ..., R_n used in an explanatory game A move the explainer closer to achieving the goal of, say, empirical fit than they were before? Then progress relative to that goal state has occurred—if not, then not.[83] And the same is the case with the rules of inference and the rules of scope.

PHILIP: But this would mean that a specific set of rules could be progressive with respect to one value, but regressive with respect to another!

STUDENT: Precisely. In short, judgments about progress are possible for different kinds of rules with respect to different values within the premises of *one explanatory game*. Take the example of the explanation of the value of goods that we have been talking about earlier. Consider the rules of representation and recall that the classical political economists of the 18th century, Smith, Ricardo and Mill, used solely natural language and numerical examples as means of representation. The marginal economists of the end of 19th century introduced the use of mathematical calculus as a new means of representation. With respect to accuracy, the mathematical calculus should be judged as superior vis-à-vis numerical examples and natural language and we have a clear case of progress here.

PHILIP: And judgments about progress made across different explanatory games?

STUDENT: It is equally unproblematic to diagnose the progress or regress of one set of rules A (i.e., of representation *or* of inference *or* of scope) used in an explanatory game X in comparison with a set of rules B (i.e. of representation *or* of inference *or* of scope) used in an explanatory game Y with respect to a specific value.

PHILIP: Give me a concrete example!

STUDENT: Consider the case of the emergence of the universe that we were discussing earlier. All three explanatory games, the biblical, the mythical and the scientific, are built around the same explanandum, though they differ radically of course with respect to the other constitutive rules, i.e. the metaphysical presuppositions and the rules that determine what must be taken as given. Now take again the case of the rules of representation. In the mythical and biblical games they consist in narratives expressed in natural language, whereas in the scientific game they consist in differential equations and computer simulations. With respect to accuracy, these last ones are certainly to be judged as superior vis-à-vis the narratives used in the mythical and biblical game.

PHILIP: However, with respect to simplicity, narratives are to be judged as vastly superior.

STUDENT: I agree. Take now the rules of inference used in the respective games. Here a comparison with respect to consistency is interesting. In both the mythical and biblical case, logical consistency among the narratives is not attained.

PHILIP: Nor is empirical consistency, since miracles are explicitly allowed!

STUDENT So, the judgment seems straightforward, that the use of logic as well as statistical and mathematical inference in the scientific explanatory game is a clear case of progress with respect to consistency. The rules of scope seem to be similar or even identical in all three explanatory games, since the domain of application of all explanations is the whole universe—no comparisons are necessary here.

PHILIP: I see how the whole approach is supposed to work. But an explanatory progress per se can only be diagnosed when all kinds of rules are taken into account, don't you agree?

STUDENT: Yes, given the multiplicity of values, such a diagnosis is more complex since it involves multi-dimensional evaluations of different types of rules. It surely is a complex judgment, but not an impossible one.

PHILIP: And as long as the broader institutional framework in which the respective explanatory games are embedded will allow for correcting the errors that the participants are always going to make...

STUDENT: Institutions are of fundamental importance. But let me state a final thought on that issue, because we are approaching the surprise that I have promised to you. So, given what I have been suggesting, there are two kinds of explanatory progress among which one has to distinguish: *local explanatory progress*, when we make judgments with respect to one explanatory game; and *global explanatory progress* when we make judgments across different explanatory games. A case of local explanatory progress can be diagnosed whenever there is a simultaneous change of all rules with respect to all relevant values adopted in the discussion. This will be a clear-cut case, but a rare one. The more frequent case will be of a variation between progress and regress across different dimensions—in those hard cases, the judgments will involve the weighing and ordering of the values.

PHILIP: And such judgments and informed choices are ultimately to be made by whom, by the participants of the explanatory game or the philosophers?

STUDENT: By both. There is—sadly—no privileged position here!

PHILIP: And decisions regarding global explanatory progress across different explanatory games will be even harder.

STUDENT: Indeed.

PHILIP: Impossible perhaps?

STUDENT: No, not impossible. I do not share Kuhn's view that there can only be local explanatory progress. I am not a relativist, as I said!

PHILIP: So, global explanatory progress is possible, you say. But there must be a basis for comparability among whole games—and what should this be?

STUDENT: The sole condition is that the constitutive rules determining what counts as an explanandum are the same across different explanatory games.

PHILIP: Can you provide an example?

STUDENT: Let me try this out in the case of cosmogony. If we compare the biblical and the scientific explanatory games, the rules of scope seem to be similar or identical, so that the real important evaluation concerns the rules of representation and the rules of inference. I would say that with respect to nearly all values, i.e. accuracy, consistency, fruitfulness, etc. the scientific explanatory game is superior to the biblical game. Only with respect to simplicity does the biblical game fare better. So, the complexity of the decision regarding the superiority of one or the other game is relatively high and presupposes the weighing of the importance of different values.

PHILIP: But…

STUDENT: Here is it! Can you see the ship under this glass shelter at the sea-side?

PHILIP: Yes.

STUDENT: This is the Ship of Theseus.

PHILIP: Amazing. This was entirely made of wood—it seems—when it was first built, but I cannot imagine that it is the original one. This must have been restored, I guess?

STUDENT: The Ship of Theseus stayed here for thousands of years. As time went on, some of the wooden planks started rotting away. To keep the ship intact, the rotting planks were replaced with new planks made of the same material.

PHILIP: So, it is not the Ship of Theseus then!

STUDENT: It has changed, but it is still the same. But we can talk about the problem of identity and constitution later. Let us head back to the top of the hill to enjoy the sunset.

Notes

1. Hempel and Oppenheim (1948).
2. According to the Tübingen School Plato's unwritten doctrines were only orally communicated to the more advanced students in the Academy and were not fixed in his writings.
3. Popper (1934). See also Popper (1949), the article with the title 'Naturgesetze und theoretische Systeme', which was a lecture given at the European Forum of the Austrian College in Alpbach, Tyrol in August 1948.
4. Goodman (1955/1983).
5. Russell (1912/1994, p. 173).
6. Bromberger (1966).
7. Kyburg (1965).
8. Salmon (1970).
9. Garfinkel (1981, Chap.1).
10. Dretske (1973).
11. Garfinkel (1981, p.173).
12. Van Fraassen (1980, Chap. 5).
13. Lewis (1986a).
14. Hempel (1965, p. 349).
15. Lewis (1986a, p. 237).
16. Salmon (1984, p. 16ff.), Wright (2015).
17. Salmon (1984, p. 131f.)
18. Salmon (1984, p. 139ff.)
19. Salmon (1984, p. 153).
20. Dowe (2000).
21. Hume (1740/1978, 1748/1975).
22. Baker (2012), Lange (2017).
23. Friedman (1974).
24. Khalifa (2012, 2017), Wilkenfeld (2014), Smith (2014).

© The Author(s), under exclusive license to Springer Nature Switzerland AG 2018   49
C. Mantzavinos, *A Dialogue on Explanation*, SpringerBriefs in Philosophy,
https://doi.org/10.1007/978-3-030-05834-0

25. Kuhn (1962/1970, p. 189ff.).
26. Darwin (1859/1967), McLaughlin (2001).
27. Machamer et al. (2000, p. 3).
28. Woodward (2003, 2015).
29. Lewis (1986b).
30. Benacerraf (1973).
31. Woodward (2003, p. 122).
32. Mill (1843/1974, Book III, Chaps. 8–10).
33. Mackie (1980, Chap. 3).
34. Strevens (2008).
35. Jackson and Pettit (1992).
36. Fodor (1974).
37. Boudon (1974).
38. Sokal and Bricmont (1997).
39. Nelson and Winter (1982).
40. Smith (1776/1976, p. 33).
41. Ricardo (1817/1951, p. 11).
42. Smith (1776/1976, Book I, Chap. VI, p. 53).
43. Ricardo (1817/1951, p. 17).
44. Malthus (1814, p. 12).
45. Mill (1848/1909, Sect. III. 4.8.)
46. Walras (1874).
47. Jevons (1871).
48. Menger (1871).
49. De Usu Pulsum (V, 155–6) and English translation by Furley and Wilkie (1984, p. 200).
50. Hankinson (2008, p. 8).
51. *De Usu Partium* III 497, =i 362, 19–123 Helmreich; more cautiously *De Anatomicis Administrationibus* II 623, 2–3.
52. *De Sectis ad eos qui Introductur*, English translation by Michael Frede *On Sects for Beginners* (1985).
53. Nutton (2008, p. 363).
54. Snellen (1984, p. 21).
55. Lindberg (2007, pp. 246ff.).
56. Butterfield (1957, p. 59).
57. O' Malley (1970, p. 4).
58. Vesalius (1543).
59. Colombo (1559).
60. Cesalpino (1571).
61. Harvey (1628/1989).
62. Malpighi (1661/1929).
63. Servetus (1553/2008).
64. Hesiod, Theogony, English Translation, 1976.
65. Bible, Old Testament.
66. Railton (1981, pp. 240ff).

67. Popper (1999).
68. Laudan (1977, p. 16).
69. Kuhn (1977, p. 321f.), McMullin (2008).
70. McMullin (1983/2012), Carrier (2013), Elliott and McKaughan (2014).
71. Neurath (1932/1933, p. 206).
72. Albert (1968/1985, p. 18f.)
73. Sextus Empiricus, Outlines of Pyrrhonism, Book 1, Chap. XV.
74. Arabatzis (2006).
75. Albert (1982, p. 40).
76. Schupbach and Sprenger (2011), Crupi and Tentori (2012).
77. Knight (1921).
78. Albert (1987, pp. 87ff.)
79. Popper (1958/1989, pp. 148ff.)
80. Jones (2003), Bernholz et al. (1998).
81. Popper (1970, p. 56f.)
82. Chang (2012).
83. Laudan (1984, pp. 65ff.)

Bibliography

Albert, Hans. 1968/1985. *Treatise on Critical Reason*. Princeton: Princeton University Press.

Albert, Hans. 1982. *Die Wissenschaft und die Fehlbarkeit der Vernunft*. Tübingen: J.C.B. Mohr (Paul Siebeck).

Albert, Hans. 1987. *Kritik der reinen Erkenntnislehre*. Tübingen: J.C.B. Mohr (Paul Siebeck).

Arabatzis, Theodore. 2006. On the Inextricability of the Context of Discovery and the Context of Justification. In *Revisiting Discovery and Justification*, ed. Jutta Schickore and Friedrich Steinle, 215–230. Dordrecht: Springer.

Aristotle. 1984. *The Complete Works of Aristotle*, ed. Jonathan Barnes, 2 Volumes. Princeton, N.J.: Princeton University Press.

Baker, Alan. 2012. Science-Driven Mathematical Explanations. *Mind* 121: 243–267.

Barker, Gillian, and Philip Stuart. 2014. *Philosophy of Science. A New Introduction*. Oxford: Oxford University Press.

Benacerraf, Paul. 1973. Mathematical Truth. *Journal of Philosophy* 70: 661–679.

Bernholz, Peter, Manfred E. Streit, and Roland Vaubel. 1998. *Political Competition, Innovation and Growth*. Berlin and New York: Springer.

Bible, Old Testament.

Boudon, Raymond. 1974. *Education, Opportunity and Social Inequlity*. New York: Wiley.

Bromberger, Sylvain. 1966. Why-Questions. In *Mind and Cosmos*, ed. Robert Colodny, 86–111. Pittsburgh: University of Pittsburgh Press.

Butterfield, Herbert. 1957. *The Origins of Modern Science 1300–1800*, rev ed. New York: Free Press.

Carrier, Martin. 2013. Values and Objectivity in Science: Value-Ladeness, Pluralism and the Epistemic Attitude. *Science & Education* 22: 2547–2568.

Cesalpino, Andrea. 1571. *Peripateticarum Questionum*. Libri Quinque, Venice: Apud Iuntas.

Chang, Hasok. 2012. *Is Water H₂O? Evidence, Realism and Pluralism*. Berlin and New York: Springer.

Colombo, Realdo. 1559. *De Re Anatomica*. Libri XV, Venice: Nicolai Beuilacquae.

Crupi, Vincenzo, and Tentori Katya. 2012. A Second Look at the Logic of Explanatory Power (with Two Novel Representation Theorems). *Philosophy of Science* 79: 365–385.

Darwin, Charles. (1859/1967). *On the Origin of Species by Natural Selection*. London: John Murray. (Facsimile reprint 1967, with introduction by Ernst Mayr, Cambridge, MA: Harvard University Press).

Dretske, Fred. 1973. Contrastive Statements. *Philosophical Review* 82: 411–437.

Dowe, Phil. 2000. *Physical Causation*. Cambridge: Cambridge University Press.

© The Author(s), under exclusive license to Springer Nature Switzerland AG 2018 53
C. Mantzavinos, *A Dialogue on Explanation*, SpringerBriefs in Philosophy,
https://doi.org/10.1007/978-3-030-05834-0

Elliott, K.C., and D.J. McKaughan. 2014. Nonepistemic Values and the Multiple Goals of Science. *Philosophy of Science* 81: 1–21.

Fodor, Jerry. 1974. Special Sciences: Or the Disunity of Science as a Working Hypothesis. *Synthese* 28: 77–115.

Frede, Michael (ed.). 1985. *Galen: Three Treatises on the Nature of Science. On the Sects for Beginners. An Outline of Empiricism. On Medical Experience*. Indianapolis: Hackett Publishing.

Friedman, Michael. 1974. Explanation and Scientific Understanding. *Journal of Philosophy* 71: 5–19.

Furley, David J., and J.S. Wilkie. 1984. *Galen on Respiration and the Arteries*. New Jersey: Princeton University Press.

Garfinkel, Alan. 1981. *Forms of Explanation*. New Haven and London: Yale University Press.

Goodman, Nelson. 1955/1983. *Fact, Fiction, and Forecast*, 4th ed. Cambridge, MA: Harvard University Press.

Hankinson, Robert J. 2008. The Man and His Work. In *The Cambridge Companion to Galen*, ed. Robert J. Hankinson, 1–33. Cambridge: Cambridge University Press.

Harvey, William. 1628/1989. *Exercitatio Anatomica de Motu Cordis et Sanguinis in Animalibus*, Engl. translation by Robert Willis in *The Works of William Harvey*. Philadelphia: University of Pennsylvania Press.

Helmreich, G. 1907–1909. *Galeni. De Usu Partium Libri XVII*. Leipzig: Teubner.

Hempel, Carl G., and Paul Oppenheim. 1948. Studies in the Logic of Explanation. *Philosophy of Science* 15: 135–175. (Reprinted in Hempel (1965)).

Hempel, Carl G. 1965. *Aspects of Scientific Explanation and Other Essays in the Philosophy of Science*. New York: The Free Press.

Hesiod. 1976. *Theogony*. London: Penguin Classics.

Hume, David. 1740/1978. *A Treatise of Human Nature*. Oxford: Oxford University Press.

Hume, David. 1748/1975. *An Enquiry Concerning Human Understanding*. Oxford: Oxford University Press.

Jackson, Frank, and Philip Pettit. 1992. In Defense of Explanatory Ecumenism. *Economics and Philosophy* 8: 1–21.

Jevons, William Stanley. 1871. *The Theory of Political Economy*. London and New York: MacMillan & Co.

Jones, Eric L. 2003. *The European Miracle*, 3rd ed. Cambridge: Cambridge University Press.

Khalifa, Kareem. 2012. Inaugurating Understanding or Repackaging Explanation? *Philosophy of Science* 79: 15–37.

Khalifa, Kareem. 2017. *Understanding, Explanation, and Scientific Knowledge*. Cambridge: Cambridge University Press.

Kitcher, Philip. 1976. Explanation, Conjunction, and Unification. *Journal of Philosophy* 73: 207–212.

Kitcher, Philip. 1981. Explanatory Unification. *Philosophy of Science* 48: 251–281.

Kitcher, Philip. 1985. Two Approaches to Explanation. *Journal of Philosophy* 82: 632–639.

Kitcher, Philip. 1989. Explanatory Unification and the Causal Structure of the World. In *Scientific Explanation*, vol. 13 of Minnesota Studies in the Philosophy of Science, ed. Philip Kitcher and Wesley Salmon, 410–505.

Kitcher, Philip. 1993. *The Advancement of Science*. Oxford: Oxford University Press.

Kitcher, Philip. 2001. *Science, Truth, and Democracy*. Oxford: Oxford University Press.

Kitcher, Philip. 2011. *Science in a Democratic Society*. New York: Prometheus Books.

Kitcher, Philip, and Salmon Wesley. 1987. Van Fraassen on Explanation. *Journal of Philosophy* 84: 315–330.

Knight, Frank. 1921. *Risk, Uncertainty and Profit*. Chicago: The University of Chicago Press.

Kuhn, Thomas. 1962/1970. *The Structure of Scientific Revolutions*, second enlarged edition. Chicago: Chicago University Press.

Kuhn, Thomas. 1977. Objectivity, Value Judgment, and Theory Choice. In *The Essential Tension*, ed. Thomas Kuhn, 320–339. Chicago: The University of Chicago Press.

Kyburg, Henry E. 1965. Comment. *Philosophy of Science* 32: 147–151.

Lange, Marc. 2017. *Because Without Cause. Non-Causal Explanations in Science and Mathematics*. Oxford: Oxford University Press.

Laudan, Larry. 1977. *Progress and Its Problems. Towards a Theory of Scientific Growth*. Berkeley, Los Angeles and London: University of California Press.

Laudan, Larry. 1984. *Science and Values*. Berkeley: University of California Press.

Lewis, David. 1986a. Causal Explanation. In *Philosophical Papers*, vol. 2, 214–240. Oxford: Oxford University Press.

Lewis, David. 1986b. Causation with Postscripts. In *Philosophical Papers*, vol. 2, 159–213. Oxford: Oxford University Press.

Lindberg, David C. 2007. *The Beginnings of Western Science*, 2nd ed. Chicago: The University of Chicago Press.

Machamer, Peter, Lindley Darden, and Carl Craver. 2000. Thinking about Mechanisms. *Philosophy of Science* 67: 1–25.

Mackie, John. 1980. *The Cement of the Universe*. Oxford: Oxford University Press.

Malpighi, Marcello. 1661/1929. *De Pulmonibus epistolae II ad Borellium*, translated into English by Young, James: Malpighi's "De Pulmonibus". In *Proceedings of the Royal Society of Medicine*, vol. 23, 1–11.

Malthus, Thomas Robert. 1814. *Observations on the Effects of Corn Laws on the Agriculture and General Wealth of the Country*. London: J. Johnson & Co.

Mantzavinos, C. 1994. *Wettbewerbstheorie*. Berlin: Duncker & Humblot.

Mantzavinos, C. 2001. *Individuals, Institutions and Markets*. Cambridge: Cambridge University Press.

Mantzavinos, C. 2005. *Naturalistic Hermeneutics*. Cambridge: Cambridge University Press.

Mantzavinos, C. (ed.). 2009. *Philosophy of the Social Sciences*. Cambridge: Cambridge University Press.

Mantzavinos, C. 2012. Explanation of *Meaningful* Actions. *Philosophy of the Social Sciences* 42: 224–238.

Mantzavinos, C. 2013. Explanatory Games. *Journal of Philosophy* CX: 606–632.

Mantzavinos, C. 2015. Scientific Explanation. *International Encyclopedia of Social and Behavioral Sciences*, 2nd ed., 302–308.

Mantzavinos, C. 2015. *Explanatory Pluralism*. Cambridge: Cambridge University Press.

Mantzavinos, C. 2016. Hermeneutics. *Stanford Encyclopedia of Philosophy*.

Mantzavinos, C., North Douglass, and Syed Shariq. 2004. Learning, Institutions and Economic Performance. *Perspectives on Politics* 2: 75–84.

McLaughlin, Peter. 2001. *What Functions Explain. Functional Explanation and Self-Reproducing Systems*. Cambridge: Cambridge University Press.

McMullin, Ernan. 1983/2012. Values in Science. *PSA: Proceedings of the Biennial Meeting of the Philosophy of Science Association*, vol. 182, vol. two: Symposia and Invited Papers, and reprinted in *Zygon*, vol. 47, 686–709.

McMullin, Ernan. 2008. The Virtues of a Good Theory. *The Routledge Companion to Philosophy of Science*, ed. Stathis Psillos and Martin Curd, 498–508. London and New York: Routledge.

Menger, Carl. 1871. *Grundsätze der Volkswirtschaftslehre*. Wien: Wilhelm Braumüller.

Mill, John Stuart. 1843/1974. *A System of Logic. Ratiocinative and Inductive*. Collected Works of John Stuart Mill, vol. VII, ed. J.M. Robson and R.F. McRae. Toronto: University of Toronto Press and Routledge & Kegan Paul.

Mill, John Stuart. 1848/1909. *Principles of Political Economy with some of their Applications to Social Philosophy*, 7th ed. London: Longmans Green and Co.

Nelson, Richard, and Sidney Winter. 1982. *An Evolutionary Theory of Economic Change*: Cambridge, MA: Harvard University Press.

Nutton, Vivian. 2008. The Fortunes of Galen. In *The Cambridge Companion to Galen*, ed. Robert J. Hankinson, 355–390. Cambridge: Cambridge University Press.

Otto Neurath, (1932/1933) Protokollsätze. *Erkenntnis* 3 (1): 204–214.

O'Malley, Charles Donald. 1970. The Lure of Padua. *Medical History* 14: 1–9.

Popper, Karl R. 1934. *Logik der Forschung*. Vienna: Springer. Imprint 1935, actually published 1934.

Popper, Karl R. 1934. *Logik der Forschung*. Vienna: Springer. Imprint 1935, actually published 1934.

Popper, Karl R. 1949. Naturgesetze und theoretische Systeme. In *Gesetze und Wirklichkeit*, ed. Simon Moser, 43–60. Innsbruck: Tyrolia-Verlag.

Popper, Karl R. 1958/1989. Back to the Presocratics. In *Proceedings of the Aristotelian Society*, N.S. vol. 59. (Reprinted in: *Conjectures and Refutations. The Growth of Scientific Knowledge*, 5th revised ed., London: Routledge, pp. 136–165).

Popper, Karl R. 1970. Normal Science and Its Dangers. *Criticism and the Growth of Knowledge*, ed. Imre Lakatos and Alan Musgrave, 51–58. Cambridge: Cambridge University Press.

Popper, Karl R. 1999. *All Life is Problem Solving*. London and New York: Routledge.

Railton, Peter. 1981. Probability, Explanation and Information. *Synthese* 48: 233–256.

Ricardo, David. 1817/1951. *On the Principles of Political Economy and Taxation*, ed. Pierro Sraffa. Cambridge: Cambridge University Press.

Russell, Bertrand. 1912/1994. On the Notion of Cause. In *Proceedings of the Aristotelian Society New Series*, vol. 13, pp. 1–26. (Reprinted in: *Mysticism and Logic*, London and New York: Routledge).

Salmon, Wesley. 1970. Statistical Explanation. In *The Nature and Function of Scientific Theories*, ed. Robert G. Colodny, 173–231. Pittsburgh: University of Pittsburgh Press. (Reprinted in Wesley Salmon, *Statistical Explanation and Statistical Relevance*, (Pittsburgh: University of Pittsburgh Press, 1971), pp. 29–87).

Salmon, Wesley. 1984. *Scientific Explanation and the Causal Structure of the World*. Princeton, NJ: Princeton University Press.

Schupbach, Jonah N., and Sprenger Jan. 2011. The Logic of Explanatory Power. *Philosophy of Science* 78: 105–127.

Servetus, Miguel. 1553/2008. *Christianismi Restitutio*, Vienne, Engl. translation: *The Restoration of Christianity*, by Christopher A. Hoffmann and Marian Hiller. Lewiston, NY: The Edwin Mellen Press.

Sextus, Empiricus. 1933. *Outlines of Pyrrhonism*, trans. R.G. Bury, Loeb Classical Library. Cambridge, MA: Harvard University Press.

Smith, Adam. 1776/1976. *An Inquiry into the Nature and Causes of the Wealth of Nations*, ed. Edwin Cannan. Chicago: The University of Chicago Press.

Smith, Ryan. 2014. Explanation, Understanding, and Control. *Synthese* 191: 4169–4200.

Snellen, H.A. 1984. *History of Cardiology*. Rotterdam: Donker Academic Publications.

Sokal, Alan, and Jean Bricmont. 1997. *Les Impostures Intellectuelles*. Paris: Editions Odile Jacob.

Strevens, Michael. 2008. *Depth. An Account of Scientific Explanation*. Cambridge, MA: Harvard University Press.

van Fraassen, Bas C. 1980. *The Scientific Image*. Oxford: Clarendon Press.

Vesalius, Andreas. 1543. *De Humani Corporis Fabrica Libri Septem*. Basel: Johannes Oporinus, English Translation: *On the Fabric of the Human Body*, by William Frank Richardson and John Burd Carman, Volume 5 containing *Books VI: The Heart and Associated Organs and Book VI: The Brain*. Novato CA: Norman Anatomy Series, No 5.

Walras, Leon. 1874. *Éléments d' Économie Politique Pure, ou Théorie de la Richesse Sociale*. Lausanne: L. Corbaz & Cie.

Wilkenfeld, Daniel. 2014. Functional Explaining. A New Approach to the Philosophy of Explanation. *Synthese* 191: 3367–3391.

Woodward, James. 2003. *Making Things Happen*. Oxford: Oxford University Press.

Woodward, James. 2015. Interventionism and Causal Exclusion. *Philosophy and Phenomenological Research* 91: 303–347.

Wright, Cory. 2015. The Ontic Conception of Scientific Explanation. *Studies in History and Philosophy of Science* 54: 20–30.